WirelessHART™

Filter Design for Industrial Wireless Networked Control Systems

T0321096

WirelessHART™

Filter Design for Industrial Wireless Networked Control Systems

Tran Duc Chung • Rosdiazli Ibrahim
Vijanth Sagayan Asirvadam
Nordin Saad • Sabo Miya Hassan

CRC Press
Taylor & Francis Group
Boca Raton London New York

CRC Press is an imprint of the
Taylor & Francis Group, an **informa** business

MATLAB® is a trademark of The MathWorks, Inc. and is used with permission. The MathWorks does not warrant the accuracy of the text or exercises in this book. This book's use or discussion of MATLAB® software or related products does not constitute endorsement or sponsorship by The MathWorks of a particular pedagogical approach or particular use of the MATLAB® software.

CRC Press
Taylor & Francis Group
6000 Broken Sound Parkway NW, Suite 300
Boca Raton, FL 33487-2742

First issue in paperback 2020

© 2018 by Taylor & Francis Group, LLC
CRC Press is an imprint of Taylor & Francis Group, an Informa business

No claim to original U.S. Government works
Version Date: 20171025

ISBN 13: 978-0-367-65751-2 (pbk)
ISBN 13: 978-1-138-29924-5 (hbk)

Visit the Taylor & Francis Web site at
http://www.taylorandfrancis.com

and the CRC Press Web site at
http://www.crcpress.com

To my beloved wife and best friend, Lidia, and my princesses Azra, Auni, Ahna and Ayla, my love will always be with you.
Rosdiazli Ibrahim

I would like to thank my parents, my wife Leah and my son's Ahil and Ashanth for their unfailing support.
Vijanth Sagayan Asirvadam

For my much-loved family.
Nordin B. Saad

To my beloved family, respective supervisors, friendly colleagues and supportive teammates who have been with me through pursuing my Ph.D. degree and completing this book.
Tran Duc Chung

To my beloved family and humanity.
Sabo Miya Hassan

Contents

Preface

Industrial wired communication protocols such as highway addressable remote transducer (HART), Foundation Fieldbus, and process field bus (PROFIBUS) have been in industries for more than 25 years. Recent advances in wireless technology have led to the emergence of international industrial wireless standards such as wireless HART (WirelessHART) and ISA100 Wireless (formerly known as ISA100.11a) that are specifically designed for industrial monitoring and control applications. The introduction of these technologies has allowed the process industries to move forward in applying them to solve some of the critical issues associated with the wired protocols such as high operation cost and low installation flexibility. At present, their main application is for monitoring process parameters while several attempts are being made for control. This is mainly because of the two major technical challenges: network induced delays and packet dropout (loss) which contribute significantly to the degradation of control performance of wireless process plants.

In order to address these issues, model-based approaches using Smith predictor and Kalman filter have been considered. However, they are complex, model-dependent, and not computationally efficient for use in low processing capability microcontroller-based field actuators. In an industrial environment, long battery life is required for these actuators, and the model-based techniques cannot satisfy this requirement.

Therefore, to tackle these two challenges, this book aims to introduce an advanced dual purpose exponentially weighted moving average (dpEWMA) filter which is simple, model-independent, and more computationally efficient than traditional filters. The filter can be implemented in any type of microcontroller without severely affecting its energy consumption. In addition, by applying the designed filter, the negative effects of network induced delays as well as packet dropouts on plant control performance are compensated. Being generic is another advantage of the filter; thus it can be applied to address the issues across industrial networked control systems, regardless of communication medium such as wires or wireless. In addition, this book also details the development of WirelessHART hardware-in-the-loop simulator (WH-HILS) that is suitable for use as a scalable validation platform for any wireless control strategies using the industrial WirelessHART technology.

This book will be of particular interest to process control engineers, especially those working towards migrating their plants from a wired to a wireless network. Other possible beneficiaries of the content of this book are researchers and students studying delay and packet dropouts compensation in a wireless networked control environment. This is specifically made easy by the use of WH-HILS.

With a total of 11 chapters, the book is structured in such a way that sequential flow is maintained. Thus the readers of the book will find it interesting since each chapter builds on the preceding chapter. For example, Chapters 1 and 2 give the introduction to general wireless technologies and WirelessHART. This includes the recent work being done in both simulation and practical environments. Chapters 3 and 4 discuss the challenges of the uncertain nature of wireless networks including delay and packet dropout and approaches for measurement of these entities. Next, Chapter 5 discusses in detail the essential delay and packet dropout compensation techniques. Furthermore, Chapters 6 and 7 cover the fundamentals of EWMA filters and advanced design of the dpEWMA filter for addressing the negative effects of both delay and packet dropout in wireless networked control systems. The remaining chapters (8, 9, 10) focus on the implementation of the designed filter for both wired and wireless industrial-like control environments and assessing its performance. Finally Chapter 11 presents some interesting directions for future investigations and continuation of the presented works.

List of Figures

List of Tables

1

Introduction

1.1 Introduction

Industrial networked control systems (NCSs) have experienced continuous technology revolutions over the past decades. One of the significant changes for process industries is moving from wired to wireless technologies, from analog to digital instruments, from pure digital instruments to smart instruments in which self-diagnostic functionalities are supported. As illustrated in Figure 1.1, the advancement of these technologies has changed control strategy from electronic single control loop in the 1960s, to single loop digital controller in the 1970s, then multi-loop digital controller for single process plants in the 1980s, and the wireless digital controller nowadays [1–3].

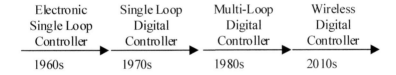

FIGURE 1.1: Revolution of wireless digital controller

For decades, industrial plants have used wires for communication with controllers [4]. The three most common open protocols for such communication are the highway addressable remote transducer (HART), Foundation Fieldbus, and PROFIBUS. HART protocol was developed in the mid 1980s by Rosemount, and in 1993 it became open protocol with its intellectual property dominated by HART Communication Foundation (now part of FieldComm Group, the result of a merger of Foundation Fieldbus and HART Communication Foundation, operating since January 1, 2015). PROFIBUS was developed in 1987 in a fieldbus project hosted by 13 companies and 5 universities in Germany [5], while Foundation Fieldbus project was started later, in 1994, by Fieldbus Foundation formed by the merging of Interoperable Systems Project and WorldFip North America [6]. Among these protocols, HART has been on track for more than 25 years with more than 30 million devices worldwide [7] and at present, over 300 member organizations to govern the development of the protocol [8]. Many of them are the worldwide leaders in process automation industries such as Emerson, Yokogawa Electric Corporation, Pepperl+Fuchs, Honeywell International, etc. This makes HART the dominant communication protocol being used in industry. With ultimate advantages for process plants, in 2007, HART protocol was revised to incorporate wireless communication capability [9]. Since then, it is known as WirelessHART. In 2010, the standard received international recognition through approval from the International Electro-technical Commission (IEC) [10]. Thus, it is the first international standard (IEC 62591) for industrial process plant control and monitoring applications.

Different from wireless technologies for home and office applications (Wi-Fi [11], ZigBee [12], mobile [13], Z-wave [14]), WirelessHART is internationally recognized as a suitable standard for industrial process control and monitoring applications. It is a field-proven technology with more than 100,000 instruments installed over 8,000 industrial networks [15]. In addition, it supports mesh topology and channel hopping to ensure that the wireless link's reliability exceeds 99%.

Even with this high reliability, WirelessHART networked control system (WHNCS), shown in Figure 1.2, faces some technical challenges from a control perspective. The first issue is the reliability of the wired link between controller or plant actuator and wireless nodes [2] which has not been addressed by any of the recent literature [11,12,16–21]. The issue can be extended to a more general problem in wireless environment, not only limited to the aforementioned wired links. The second issue is network induced delay which affects arrival time of a wireless packet to its destination. It is interrelated to packet dropout (loss). In a WNCS, a controller is scheduled to perform control action at a short given specific time (sampling time). A delayed packet arrival which goes beyond the sampling time is defined as packet dropout. Taking this phenomenon into consideration, the occurrence of packet dropout can happen at both wired links and wireless links. Therefore, the major contribution of this book is to discuss in detail an approach using advanced EWMA filter design technique to address these issues for improvement of the overall control performance of the system.

1.2   Communication Technology Evolution in the Industrial Process

In industrial process plants, the communication technology evolves from simple wired communication (using analog signals, then digital signals), to wireless communication (which consists of the transformation of signals: (i) from

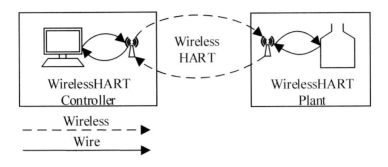

FIGURE 1.2: WirelessHART process plant with wired and wireless links

digital to analog at transmitter, then (ii) from analog back to digital at receiver). At present, analog communications in industrial environments do exist; however, they play a minor role compared to digital communications. The digital communication using wires is still widely used in most process plants worldwide. The communication is dominated by some common protocols namely PROFIBUS, HART, Fieldbus, controller area network (CAN) and local area network (LAN). Recently, massive interest has been placed on transforming digital wired plants to wireless plants. The simplest application found to be common is the use of Wi-Fi routers for wireless data transmission between clients and servers. To some extent, ZigBee is also being used; however, due to its limitations such as being prone to error and low speed (compared to Wi-Fi), it is mainly used in non-critical applications such as in farming industries. Wireless technologies like WirelessHART, ISA100 Wireless are becoming dominant in industrial process plants. This is because they are designed and dedicated for applications in harsh environments with special features to support channel hopping, mesh networking, and channel blacklisting for eliminating blocking channels in industrial plants.

It is seen that the plant industries are now moving toward wireless operation, firstly through the introduction of wireless monitoring devices which allow remote monitoring of multiple process parameters at a reasonable data update rate. This not only helps to simplify the monitoring process, but also to reduce maintenance cost associated with a large chunk of cables across the plant field.

1.3 Industrial Internet of Things and the Demand for Wireless Control in the Process Industry

Recent advancement in wireless technologies and miniaturization of electronic circuits have led to the development of a wide range of devices that are able to stay connected with both local wireless network and Internet. This results in a hot research topic emerging in recent years, namely the Internet of Things focusing overall architecture [22, 23], cross layer protocol [24], security [25], energy efficient systems [26] etc. [27, 28]. Although IoT is currently applied mainly in home and office applications, soon its application in industrial environment will become significant. Therefore, research on the Industrial Internet of Things (IIoT) has been receiving substantial attention from both academic and industrial communities worldwide [29,30]. With IoT, any electronic equipment or even a single component installed at industrial process plants will be able to transmit and provide its own information to the local network, then even across locations via Internet connections. This is considered a strong tool to help plant management understand conditions in their plants in a second or less. Thus, monitoring of critical components in the process plants has become easier than before. In addition, information on the equipment becomes available anytime, anywhere especially with the support of cloud technology. This is not only improving the plant's monitoring activities, but also smoothing maintenance activities. Much information on plant equipment allows maintenance to be scheduled proactively through trending analysis and early fault detection technologies supported by big data mining tools. In order to make information ready and avoid tremendous amounts of messy cables, wireless technologies for industrial communication are definitely needed for adoption, thus IIoT. With wireless communication, the entire plant stays connected locally and globally as shown in Figure 1.3.

As seen from the figure, the headquarters of a corporation can set up a main controller for its entire nationwide or worldwide plants. The Internet and cloud service are used to maintain connectivity between the headquarters and remote plant network backbones. The remote site connects with the Internet through a router. The network connection is protected by a firewall. At the plant site, the remote controller can be located at a main office of the plant site and is connected to the network backbone, while wireless connection is used to maintain the network between the network backbone and all production plants located in the site. The connection is maintained by a gateway and an access point. At each plant, there is a local controller that does the control tasks helping the plant to achieve its desired production output. However, it can receive control command from the remote controller as well. It also will connect with handheld devices to provide on-site information for diagnostics and maintenance activities by site engineers. Thus, overall IIoT helps to ensure each individual component of industrial process plants connects together and with global offices. Therefore, it can be seen that there is a strong demand for adoption of wireless technologies in advancing process industries.

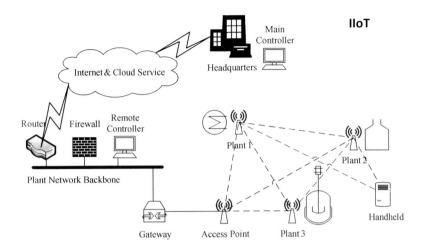

FIGURE 1.3: Industrial Internet of Things (IIoT)

1.4 This Book's Scope and Organization

Overall, this book is organized as follows.

- Chapter 1 introduces networked control systems and the evolution toward wireless networked control systems (WNCSs), an overview of Industrial Internet of Things (IIoT) and the demand for wireless technologies to be successfully applied in industrial process plants.

- Chapter 2 covers a comprehensive literature review of industrial wireless NCSs and related challenging issues. It also presents WirelessHART protocol for industrial WNCS, its applications and current industrial challenges.

- Chapter 3 highlights several uncertainties in industrial wireless mesh network that can affect the overall networked control performance.

- Chapter 4 discusses major approaches for realizing and measuring the delay and packet dropout in industrial wireless networked control systems.

- Chapter 5 outlines some important techniques to address delay and packet dropout for enhancing overall networked control performance, their advantages and disadvantages.

- Chapter 6 covers the basics needed to gear toward applying and designing exponentially weighted moving average (EWMA) filters for use in wireless networked control systems.

- Chapter 7 details advanced methods for designing a suitable dpEWMA filter for addressing both delay and packet dropout issues in wireless networked control systems. The filter's performance is analyzed fundamentally through the analysis of both setpoint tracking and disturbance rejection performances, with and without consideration of environment noises.

- Chapter 8 presents research results from the method presented in Chapter 7 with focus on addressing delay and packet dropout over wired links in wireless networked control systems.

- Chapter 9 introduces a novel platform for validation of wireless networked control system's algorithms using a hybrid software-hardware approach. Here, the design of a hardware-in-the-loop WirelessHART simulator, its usage and potential applications are presented in detail.

- Chapter 10 analyzes the filter's performance in a real-time industrial-scaled mesh network. The results include the measurements of delay and packet dropout over wired link, wireless network statistics and induced delay, control performance of plants with different orders using the proposed EWMA filters as compared to Smith predictor and Kalman filter, and experiment results with the developed real-time WirelessHART hybrid simulator (WH-HILS).

- Finally, Chapter 11 concludes the findings and contributions of this book and its significance as well as potential applications in industrial environments.

2

WirelessHART, The Leading Technology for Industrial Wireless Networked Control Systems

2.1 Introduction

Networked control systems (NCSs) can communicate mainly through wires (including fiber optic) or wirelessly. Several widely used wired protocols are controller area network (CAN), Ethernet, Modbus, Foundation Fieldbus, PROFIBUS, and HART. The cost, time and large number of cables for installation and long maintenance time are key drawbacks of wired-NCS [31, 32]. In mobile operation where flexible installation and rapid deployment are required, wireless communication is preferred. The common home and office wireless protocols are Bluetooth [33], ZigBee [34], wireless local area network (WLAN) or Wi-Fi [11]. However, they are not suitable for industrial applications due to stringent requirements such as high reliability, accuracy, timeliness and losslessness of the wireless communication. The only two standards that satisfy these requirements are WirelessHART [10] and ISA100 Wireless [35]. However, at present, the applications are limited to monitoring purposes [18, 21, 36, 37], and only few attempts on controlling applications [38–41].

In general, NCS can be categorized into two categories: wired NCS and wireless NCS (WNCS) which correspond to their communication mediums. For clarification purposes, as shown in Figure 2.1, WNCS can be further divided into two categories, home and office WNCS and industrial WNCS. The typical home and office WNCS utilizes Wi-Fi, ZigBee, Bluetooth, and Z-wave technologies while the industrial WNCS are well known with WirelessHART and ISA100 Wireless technologies.

The following sections provide a brief introduction of wired NCS and its disadvantages and detailed introduction of the counterpart, WNCS and its related technologies.

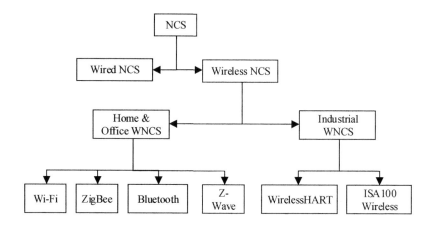

FIGURE 2.1: NCS branches

2.1.1   An Overview of Networked Control Systems and Their Classification

In a complex system like an industrial process plant, the system consists of thousands of pieces of equipment and control loops. Each control loop can operate on its own; however, in order to ensure integrity operation of the entire plant, all single control loops need to be connected and controlled by a master control room. The connections between a network of all control loops form a networked control system. Depending on the communication medium, the networked control system can be mainly categorized into wired and wireless networked control systems. The wireless category can be further divided into wireless networked control systems for home, office and industrial applications. The following sections will further discuss these types of networked control systems in detail.

2.1.2   Wired Networked Control Systems

Conventional NCS has event-triggered controllers and actuators operating in response to time-triggered sensor nodes. Hence, it is suitable for applications that need spatial distribution. The system requires a specific real-time network to ensure end-to-end latency (known as delay) and is bounded for proper operation. LAN cannot satisfy this constraint because unpredictable transmission will cause multiple retransmissions and random medium access delays. Therefore, NCS cannot operate reliably over the shared LAN network [42]. On the other hand, CAN, developed in the 1980s by Bosch GmbH, is a multicast-based protocol which was originally proposed for the automotive industries.

Later, its application expanded to plant robotics and medical industries as well. One key problem with a system using CAN is that the speed is limited by the bus length (max speed is 1 Mbps at bus length of only 30 m) and it is sensitive to the electrical characteristics of the medium (wires) [43]. Therefore, its application is usually limited to high-speed short-distance communication. In contrast, PROFIBUS is based on RS-485 physical layer. It was originally designed to replace a point-to-point wire system by a field bus system which significantly reduces the number of wire pairs in the network. Its data packet can have a large size of 244 bytes and peak speed of 12 Mbps which is higher than that of wired-HART (typically at 1.2 kbps). In addition, compared to wired-HART (the most widely used protocol for wired communication in industrial process plants), it has an interoperability advantage because wired-HART messages can be encapsulated and transmitted over the PROFIBUS system. However, like other wired protocols such as Foundation Fieldbus, the major drawback of this protocol is that it only supports trunk, star, and tree topologies which are prone to error such that if one trunk is disconnected, the entire network is affected. Furthermore, the wire length is limited depending on the number of spur lines, e.g., maximum 32 spur lines is suitable for non-intrinsically safe network with length of less than 1 m [5]. Hence, it is seen that the use of wire protocols in industrial process plants results in several limitations in type of network topology supported, maximum communication length, and maintenance cost.

2.1.3 Home and Office Wireless Networked Control Systems

The common protocols of WNCS for home and office applications are WLAN [44], Bluetooth [45], ZigBee [46], and Z-Wave [47]. WLAN offers various speed ranges from as low as 1 Mbps to as high as a few hundred Mbps; it does not support mesh network topology. Bluetooth, on the other hand, offers speeds up to 1 Mbps at low energy consumption mode for its latest version 4.2 [48]. The key disadvantage of Bluetooth is that it does not support relaying messages among slave nodes and only pico-net (small network with maximum of 7 slaves) is supported. ZigBee has low power consumption, low speed communication and offers connection of multiple devices in a network. The revision of its technology supports some network characteristics similar to the industrial protocols such as mesh network topology [34]. On the other hand, Z-Wave is technology dedicated for home and office automation which does not require very high reliability for critical control application [47].

The main issue with these protocols is their reliability. The reason is that they are prone to error and often require multiple retransmissions to improve packet delivery rate. These cause abrupt increases in network delays and degrade WNCS control performance. Furthermore, certain medium access control (MAC) parameters of these protocols greatly affect the stability of WNCS [49]. Therefore, they are not recommended for use in industrial control applications.

2.1.4   Industrial Wireless Networked Control Systems

Due to lack of an open international standard that fits the industrial requirements, WLAN, Bluetooth, ZigBee, Internet Protocol version 6 (IPv6) over low-
power wireless personal area networks (6LoWPAN) are not widely adopted
for industrial wireless applications. At present, the two industrial international
wireless standards for WNCS are WirelessHART or ISA100 Wireless [50]. Both
aim at noncritical wireless applications for control and monitoring purposes.
However, WirelessHART has been preferred by the industries since its legacy
HART protocol contributes the dominant adoption in industrial equipment.
At field level, it is expected to be used in about 80% of industry wireless communication. For these reasons, this research focuses on WirelessHART NCS
(WHNCS). The key similarities and differences between these two protocols
are presented in Table 2.1.

TABLE 2.1: Comparison of WirelessHART and ISA100 Wireless

Item	WirelessHART	ISA100 Wireless
Key Component Devices	Field device, Hand-held device, Adapter, Gateway, Network Manager, Security Manager	Input/Output (I/O), Provisioning, Router & Backbone Router, Gateway, System Manager, Security Manager, System Time Source
Field Device as Router	Yes	No
Network Topology	Mesh	Mesh
Max. Nodes	50 - 100	50 - 100
Scalability	Thousands of Devices	Thousands of Devices
Interoperability	Maximum: Estimated to Account up to 80% Wireless Communication Devices in Process Plants	Have Some Issues with Selected Features from Vendors
International Standard	From 2010	From 2014

    As seen from the table, the common components between WirelessHART
and ISA100 Wireless are wireless gateway, router, security manager, system
manager or network manager. Different from WirelessHART, in ISA100 Wireless, Input/Output (I/O) refers to a sensor or an actuator that provides or
uses data from other devices; provisioning enables other devices to join the
network. Backbone router channels data from backbone network to application devices and system time source maintains a master time source for the
whole network. For WirelessHART, normally, gateway or network manager

can hold the roles of network manager, security manager, and gateway at the same physical, embedded device. In ISA100 Wireless, backbone router is slightly different. It can have functionalities of a gateway, system manager, security manager and router. Additionally, both protocols' backbone networks are wired network connecting other wireless devices. WirelessHART field devices and adaptors can serve as relays allowing message distribution in the mesh network topology while in ISA100 Wireless, I/Os are separated from the routers [50].

2.2 WirelessHART, The Protocol and Applications

This section introduces in detail WirelessHART protocol and its features. Furthermore, a summary of state-of-the-art WirelessHART with regards to control and monitoring applications in industrial plants is presented.

2.2.1 The Protocol

WirelessHART is the first open-standard wireless technology for process industry [51,52]. It operates in 2.4 GHz industrial, scientific and medical (ISM) band using 15 different channels with the maximum data rate of 250 kbps [51]. Channel 26 is not used since some regions in the world do not allow the frequency on this channel to be operated freely. Thus, the channel index, the 15th bit, is always 0. The details of frequency channels for WirelessHART are presented in Table 2.2.

WirelessHART supports time division multiple access (TDMA), frequency hopping spread spectrum (FHSS) and direct sequence spread spectrum (DSSS) to provide fully-redundant mesh routing and offers very high reliability for data transmission (up to 99.999%). The TDMA mechanism (see Figure 2.2) is described as follows [10]. Given a time slot, both wireless source

TABLE 2.2: WirelessHART channels and frequencies

Index	No.	Frequency (MHz)	Index	No.	Frequency (MHz)
0	11	2,405	8	19	2,445
1	12	2,410	9	20	2,450
2	13	2,415	10	21	2,455
3	14	2,420	11	22	2,460
4	15	2,425	12	23	2,465
5	16	2,430	13	24	2,470
6	17	2,435	14	25	2,475
7	18	2,440	15	26	2,480

and destination must be active during this transmission time slot. First, the
transmission starts, and the message is sent from the source to the destination. When the destination receives the message, if required, it needs to send
back to the source an acknowledgment message stating whether the message
has been received successfully or not. This is to ensure reliable communication in the network. The offset time before the transmission starts is to make
sure the destination has sufficient time to start up its initialization operation
(e.g., wake up from sleep) and actively listen to the message sent from the
source. Similarly, before the acknowledgment message is sent, there is a buffer
time for the source to change its state to listening mode (being a destination).
To guarantee successful communication, both source and destination need to
know when the time slot for communication between them starts. Thus, a
network-wide time synchronization mechanism is applied to all nodes in the
network on a periodic basis. A superframe is a set of several continuous time
slots for communication between one pair of the network devices. Often, it is
used to improve communication reliability in the network.

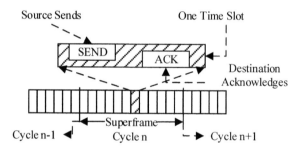

FIGURE 2.2: TDMA: one-time slot in a superframe

The mechanism of channel hopping is visualized in Figure 2.3 [10]. Communications of network devices are assigned to superframe, time slot and
communication channel (through channel offset). With channel hopping, at
the same superframe and time slot, multiple devices can communicate at different channels or frequencies. This results in multiple possible links between
any pair of devices for enhancing communication reliability over the network.
Additionally, channel blacklisting can be used to deactivate a channel that is
susceptible to noise or causes unreliable communication in the network [51].

The physical layer of WirelessHART employs the 2.4 GHz DSSS with offset
quadrature phase-shift keying (O-QPSK) modulation based on IEEE 802.15.4-
2006 standard. The transceiver specification details are shown in Table 2.3.

As described earlier, the DSSS helps spread the communication frequencies
over a range around 2,450 MHz, thus resulting in multiple channels at differ-

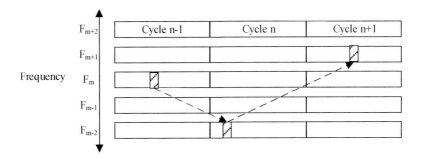

FIGURE 2.3: Channel hopping mechanism

TABLE 2.3: Transceiver specification

Freq. (MHz)	Comm. Rate (kchips/s)	Bit Rate (kb/s)	Sym. Rate (ksymbol-s/s)	Sym.
2,400 - 2,483.5	2,000	250	62.5	16-ary Orthogonal

ent frequencies for more possible network links and improved communication reliability in the network. With self-healing and self-organizing capabilities [53], and ability to support update rates of less than 1s, which is sufficient for automated process application, a WirelessHART network can support up to hundreds or thousands of devices. Furthermore, it supports star topology to maximize a device's battery lifetime and mesh topology to improve a network's reliability and coverage. By utilizing advanced encryption standard (AES) 128 encryption mechanism, WirelessHART communication is secured [51]. However, currently, its application is mainly for process, equipment and environment monitoring, asset management, and advanced diagnostics but not for control [53]. A typical WirelessHART network is shown in Figure 2.4. The network consists of a host application, a gateway serving as network manager and security manager, an access point and three field devices (Node 1 to Node 3) [20].

2.2.2 WirelessHART's Applications

Currently, the main application of WirelessHART is limited to the monitoring process while the control application is still in the exploring stage. Some recent works have reported on the simplest control strategy, on-off control, of non-

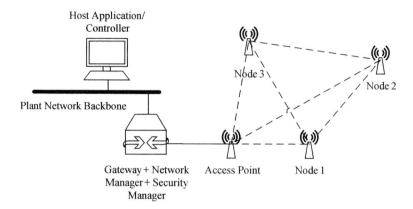

FIGURE 2.4: Typical WirelessHART network

critical process plants using the protocol. Thus, much effort is being made to bring its control application to industry level.

2.2.2.1   Monitoring Application

With the advantages in significant cable and cost reduction, simple and fast deployment, WirelessHART instruments have been implemented in many industrial process plants in recent years. The number of WirelessHART instruments has exceeded 100,000 worldwide [15] since 2012 and it is continuing to grow through the support of over 350 leading manufacturers who are members of FieldComm Group [8]. Typically, the update rate of field instruments is within the range of 4 seconds to 60 minutes [54]. This is to ensure the lifetime of field instruments between 4 and 10 years. Practically, when there is no obstruction, a WirelessHART device is placed at least 2 meters above the ground. With 10 mW or 10 dBi of power, effective communication range is 228 m. The range can be significantly reduced to as low as 30 m when the device is surrounded by heavy obstruction. As a guideline, every WirelessHART network should have minimum 5 WirelessHART devices within the effective range of the gateway. Then, every WirelessHART device should have at least 3 other devices placed in its effective range to ensure that another two connections will be used in case the main connection is broken (guaranteeing redundancy and reliability). For every network with more than 5 devices, at least 25% of the devices are within the effective range to guarantee proper bandwidth and eliminate pinch points. Particularly, a network with greater than 20% of devices with update rates faster than 2 seconds should increase the percentage of devices in the effective range of the gateway to 25% to 50%. The rule of maximum distance is that wireless devices with update rates less than 2 seconds should be within 2 times the effective range of wireless devices from the gateway [55]. This is applied to maximize speed of response for monitor and control applications requiring high-speed updates. Some concerns with monitoring applications lie in the power consumption of field instruments, the reliability of the measurements, timeliness of the measurements, and self-diagnosis capability for remote troubleshooting and maintenance scheduling [20, 56, 57].

2.2.2.2   Control Application

A network's reliability is the main concern when considering the control application of industrial wireless technologies. It is mainly associated with delay [16, 58] and packet dropout [2]. Delay is due to the capability of the system to provide timely control and feedback information. With limited power supply for field devices, the currently practical update period of a field device is limited to 4 s [54]. An update period shorter than this value will result in significant drainage of battery capacity over a short period, thus, shortening the lifetime of the field device [56]. However, depending on the nature of the application, shorter than 4 s, or even 1 s delay can be achieved with WirelessHART. In such applications, the field device should have a constant, reliable power supply. As a guideline, the update rate should be 3 to 4 times faster than process time constant for monitoring application and open loop control application. While for regulatory closed loop control and some types of supervisory control, the update rate should be 4 to 10 times faster than the time constant of the process [55, 59]. These rules are also applied when wireless feedback is used in plants. It should be noted that, in such cases, the control path is still using wires for communication [1]. The main reason is because it is not yet reliable to enable wireless control path when using a traditional controller strategy such as PID.

2.3   WirelessHART Simulators

In this section, the development history of simulators for WirelessHART is provided. Further review of the existing simulators is presented to assist readers in selecting a suitable simulator for their applications. In addition, an introduction to a WirelessHART hardware-in-the-loop simulator is discussed providing a suitable real-time, scalable simulation platform for both industrial and academic needs.

2.3.1   Simulator Development History

Figure 2.5 visualizes the development road map of WirelessHART simulators and Table 2.4 details their study topics. Overall, WirelessHART simulators can be categorized into two types: (i) legacy simulator which is based on software; and (ii) hybrid simulator which involves both hardware, i.e., WirelessHART-enabled devices, and software.

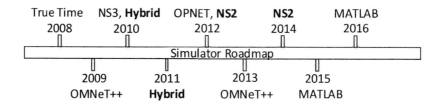

FIGURE 2.5: WirelessHART simulator roadmap

TABLE 2.4: WirelessHART simulator development history and study subjects

Year	Study Subject
2008	WirelessHART simulator using TrueTime [60], WirelessHART with clock drift and packet losses [61,62]
2009	Coexistence of WirelessHART network with IEEE 802.11b network using OMNeT++ [63]
2010	Hybrid framework for WirelessHART simulation [64], WirelessHART simulator using NS-3 [65]
2011	Evaluation of WirelessHART-enabled devices in controlled simulation environment using hybrid simulator [64]
2012	WirelessHART simulator using OPNET [66]
2013	Improve WirelessHART network simulation using OMNeT++ [51]
2014	Implement WirelessHART simulator using NS-2, correctness verification via real-world network setup [18]
2015, 2016	WirelessHART simulator using MATLAB® involving hardware for delay and packet dropout studies [2, 20, 38]

2.3.2   An Overview of Existing Simulators

In this section, an overview of the existing WirelessHART simulators is presented. It covers both software-based and hybrid (both software- and hardware-based) simulators. The section then highlights the ultimate need for a real-time hardware-in-the-loop simulator for simulation of industrial wireless process plants using the WirelessHART technology. The simulator's concept is not limited to the mentioned technology; it can be extended to other available technologies such as ISA100 Wireless as well.

2.3.2.1 TrueTime-MATLAB®-based Simulator

TrueTime is a MATLAB® and Simulink-based toolbox that is suitable for simulation of a real-time control system. It originally supports two wireless standards: IEEE 802.11b WLAN, and IEEE 802.15.4 Zigbee [60]. To study delay compensation in the WirelessHART networked control system (WHNCS), TrueTime was modified by adding additional functionalities that reflect WirelessHART characteristics, i.e., adding number of channels, slot size, super frame size into medium access control (MAC) layer, moving MAC protocol block from the wireless network block into each device [61]. Similarly, when studying packet losses of WHNCS, some WirelessHART network behaviors were introduced, i.e., MAC algorithm, simplified device communication table, and device task synchronization [61]. However, TrueTime's detail modification was not covered when studying the effect of clock drift on a closed-loop WHNCS [62]. In TrueTime, the WirelessHART communication stack was not fully implemented thus leading to incomplete support of the entire protocol.

2.3.2.2 OMNeT++-based Simulator

WirelessHART coexistence with IEEE 802.11b WLAN was studied through a simulator built based on OMNeT++ and Mobility Framework [63]. In the work, the WirelessHART communication stack consists of only three layers: Application or Network Layer, MAC Layer, and Physical Layer. To ensure communication in this dual network, channel control for each network type was implemented; this allows message propagation between the two networks' multi-nodes through a common interference module. Since the Mobility Framework toolbox (currently no longer maintained) [67] does not provide complete support to simulate IEEE 802.15.4 standard, the WirelessHART communication stack was not fully modelled. Additionally, the simulator does not differentiate between IEEE 802.15.4 suitable for low-rate wireless personal area network (LR-WPAN) and the modified IEEE 802.15.4-2006 suitable for WirelessHART in its channel control [63]. Another co-simulation approach for improving simulation of WHNCS based on WirelessHART was implemented by using OMNeT++ with UnibsFramework and TrueTime [51]. To model a process plant, the simulator requires creation of a Unix/Linux-based inter process communication (IPC) shared memory area which is accessible by both OMNeT++ and TrueTime. However, the intermediate connection, IPC, limits the simulator to be run on Unix/Linux systems only.

2.3.2.3   OPNET-based Simulator

In Figure 2.6, to study throughput and packet loss, an OPNET-based WHNCS simulator was built with only three nodes: network manager, sink node and WirelessHART device [66].

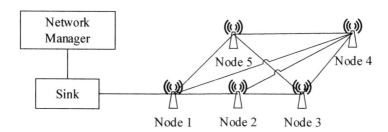

FIGURE 2.6: OPNET WirelessHART model

The sink acts as a centralized router that routes packets to all destinations. The network manager performs communication scheduling for all devices on a periodic basis. Each wireless node can be an access point to route signals and is represented by a five-layer model. The bottom layer is the Physical Layer for radio communication. It contains two processes: transmitting and receiving signal based on TDMA scheme. Above the Physical Layer is the Data Link Layer that models two processes as well: local link control (LLC) and MAC. The third layer is the Network Layer and just above it is the routing process model. The top layer contains source and sink process models. In short, the simulator requires modification of OPNET source code, and only original LLC function is utilized in the simulator. Even though communication in the model is burst mode, in a real WirelessHART network, burst mode is not often used, hence the simulator is not yet complete and it should be further developed.

2.3.2.4   NS-3-based Simulator

In NS-3-based WirelessHART simulator, only Physical Layer and a preliminary development of Link Layer were modeled [65]. Additionally, devices' roles, i.e., router, gateway, network manager, hand-held device and field device, were not differentiated in the simulator. Therefore the WirelessHART communication stack was supported incompletely by the simulator.

2.3.2.5   NS-2-based Simulator

In 2012, WirelessHART network simulator based on NS-2 was developed [68]. Recently, a more complete WirelessHART communication stack was implemented (see Figure 2.7) and the simulator's correctness was verified using an actual WirelessHART network presented in Figure 2.8 [18]. A list of tables in each communication stack's layer is outlined in Table 2.5.

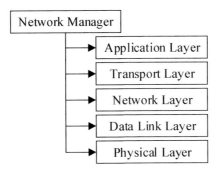

FIGURE 2.7: NS-2's full WirelessHART communication stack

TABLE 2.5: Tables in WirelessHART communication stack

Layer	Table
Transport Layer	Transport Table
Network Layer	Session Table Route Table Source Route Table Service Table
Data Link Layer	Graph Table Neighbor Table Link Table Superframe Table

In the simulator, the gateway plays the function of a network manager which schedules and manages packet routes and also security manager which secures all communication in the network. The network management algorithm has four operations: joining network, defining graph and route, scheduling communication, and service requesting. However, only the joining and service request procedures were based on WirelessHART standard.

The Physical Layer was utilized by reusing the IEEE 802.15.4 physical layer in NS-2 [18]. The communication table in the Data Link Layer was based on IEEE 802.15.4e. The IEEE 802.15.4-2003 MAC protocol [18] was modified to support: time synchronization, channel hopping, slotted unicast communication, acknowledgment at link layer and link activation. The time synchronized mesh protocol (TSMP) developed by DustNetworks and time slotted channel hopping (TSCH) were employed in the simulator. TSMP is a media

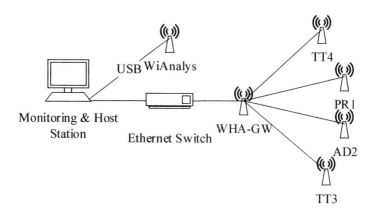

FIGURE 2.8: Actual WirelessHART network setup for simulator's correctness verification

access and network protocol specifically designed for low power, low bandwidth communication that requires reliability while TSCH is a MAC scheme and a subset of TSMP [18]. To support application in industries, IEEE 802.15.4-2006 standard was amended to introduce TSCH mode in IEEE 802.15.4e [18]. The Network Layer supports both graph routing and source routing. The Transport Layer supports transactions with both acknowledgment and without acknowledgment. Unacknowledged service is used for communication from sensor to Network Manager on best-effort basis, while acknowledged service is used for reliable communication from Network Manager to Actuator. Additionally the Transport Layer also supports a transaction involving multiple HART commands to support one-time reading (sending) several parameters from (to) a network device [18]. The Application Layer is Tcl command-based enabling interfacing between devices using commands to initialize simulation parameters, create node, and configure the simulator. It is noted that connection between Gateway and Access Point was assumed to be wireless [18], while in other simulators, the connection is assumed to be wired. However, the Access Point module ensures network scalability is supported.

Even though the simulator fully supports the WirelessHART communication stack [18], the simulator should support simulation of WHNCS. Real-time simulation capability should be added to the simulator as well.

2.3.2.6 Hybrid Simulator

To lighten the simulator's correctness verification task, a hybrid simulator (see Figure 2.9) was developed to simulate WHNCS [64].

The software used in the simulator were Contiki (Ubuntu-based [69]) oper-

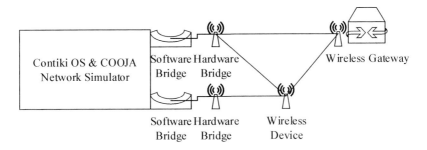

FIGURE 2.9: Hybrid simulation with COOJA network simulator

ating system and COOJA network simulator, while the hardware were Tmote Sky, SmartMesh IA-510 D2510 Network Manager, and M2510 Evaluation Mote Module. In order to enable the network simulation, the COOJA network simulator had to be modified to meet WirelessHART characteristics such as TDMA and security. In addition, both software and hardware bridges needed to be developed. The hardware bridge (Tmote Sky module) is used to interface with WirelessHART devices while the software bridge is used to interface the hardware bridge with the network simulator. Performance evaluation of the network was limited to measurements of single-hop round trip delay, multiple hop round trip delay, acknowledgment timing and time synchronization stability. In addition, not supporting the full WirelessHART communication stack is the key drawback of the hardware bridge as this causes communication incompatibility with the native network's devices due to untimely message constructions, thus network-wide time synchronization. Furthermore, the network simulator only supports network simulation which is suited for network design and characteristics studies while lacking functionalities for studying wireless control of process plants.

2.3.2.7 Summary

A summary of WirelessHART simulator bases is presented in Table 2.6. It is seen that most of the simulators are software-based and only NS-2 developed in [68] supports WirelessHART communication stacks in simulation. However, the simulator's verification had been done for only a simple network setup. In order to make the simulation more realistic, a hybrid simulation approach was proposed. However, the network's scalability and incompatibility (with native WirelessHART devices) are the main challenges for the proposed simulator. This is because it requires hardware bridges to be physically attached to the simulator's host device (computer) and the number of available ports on the computer is always limited. Thus, as the network size increases, the simulator is unable to facilitate the growing number of hardware bridges. Also, the simulator lacks the capability to support wireless networked control system (WNCS). Therefore, the following section will propose a scalable hardware-based simulator for simulation of WNCS.

TABLE 2.6: Summary of WirelessHART simulator bases

Simulator Base	Software/Hardware
TrueTime, MATLAB®	Software
OMNeT++, Mobility Framework	Software
OPNET	Software
OMNeT++, UnibsFramework, TrueTime, MATLAB®	Software
NS-2	Software verified by Hardware
NS-3	Software
Contiki, COOJA	Hybrid (Mixed)

2.3.3   Toward WirelessHART Hardware-in-the-loop Simulator

Hardware-in-the-loop (HIL) simulation is a technique often used in testing and developing complex industrial systems. It offers cost effective solutions [70, 71] for testing the complex systems [72] while helping to reduce development duration and shorten the product's time-to-market, improving safety of test personnel and expediting the validation process. Some areas which widely utilize this technique are automotive [73–75], power electronics and systems [76–79], robotics [80], and radars [81]. Since 2007, the studies on industrial WNCS using the emerging industrial wireless technology such as WirelessHART [82] are mainly for the monitoring application [16], while the control application is not yet matured and is being explored by researchers [31, 38, 50, 58, 83]. Even though based on the successful wired HART communication protocol for over 25 years, the adoption of this technology for control application is limited to the laboratory scale due to the lack of confidence in evaluation of control performance of the processes under the wireless communication environment. In order to address the issue, some simulation platforms have been proposed to enable intensive exploration of control in WirelessHART. They are mainly based on wireless simulators such as NS-2 [18], NS-3 [84], OPNET [66], OMNeT++, and TrueTime [51]. The common issue associated with these simulators is the lack of experimental validation except as reported in [18]. This is the only simulator to date that claims to fully support WirelessHART communication stack and limited features have been validated by real experimental setups. However, it has major limitations such as the validation is applicable to the plant setup presented in the corresponding paper only. Additionally it is difficult to extend the simulation validation beyond the one reported in the paper to other various types of industrial process plants.

Therefore, this research proposes a framework for development of a real-time WirelessHART-hardware-in-the-loop simulator (WH-HILS) for industrial WNCS applications. By using certified wireless instruments, the need for wireless communication stack verification in the simulator is omitted. The simulator is generic, thus can be used to simulate any industrial process plant and any type of controller. The standard communication ensures interoperability of the simulator with industrial-rated instruments such as pressure, temperature, humidity wireless transmitters. These are the key advantages of the developed simulator.

2.4   Challenges with WirelessHART

As mentioned earlier, recent works related to WirelessHART mainly focus on monitoring applications while its control application is still in the premature stage. This is due to the current control challenges of the wireless protocol when it is used for the control path. Most related works published recently focus on network induced delay and packet dropout, the two critical issues of WNCS. Due to the dynamics of most systems and the time-varying nature of operating environments, real systems always experience time-varying delay. A delayed signal, if exceeding an acceptable threshold, is considered as packet dropout. Therefore, time-varying delay and packet dropout are usually taken into consideration together in WNCS [85–88]. These issues will be discussed in the following sections. In addition, the correlation between delay and packet dropout will be presented. Energy inefficiency in wireless monitoring and control plants is another important issue that should be taken into consideration.

2.4.1   Delay Issues

In WNCS, delay is always a fundamental issue that significantly affects system performance. Wireless link delay has been studied and reported by Nawaz and Jeoti [21], Huang, et al. [39], Akerberg, et al. [89], Xiuming, et al. [90], Biasi, et al. [61], Zand, et al. [68], Ferrari, et al. [91], Saifullah, et al. [92], Remke and Xian [93] however no specific solution to address the delay has been reported. Both wireless delay and wired delay can be modeled based on statistical measurement data [4, 11]. Although the delay model helps to predict end-to-end or round-trip delay in the network, from a control system perspective, detail delay compensation was not proposed by Huang, et al. [4, 11], Nawaz and Jeoti [21]. To minimize delivery delay and packet dropout in high data rate wireless monitoring applications [12,21], WSN clock [17] and time synchronization schemes [19] have been used. The purpose is to differentiate quality of service over the network, however, they are not suitable for addressing the network induced delay and packet dropout between controller and gateway, actuator and wireless node, or between wireless nodes. To compensate for delay in the first-order stable WHNCS, proportional-integral (PI) controller design using the lambda tuning technique, digital PI controller, and digital PI controller detuned was used [62]. However, the procedure for tuning parameters of the controllers was not presented. Alternatively, predictive PI (PPI) controller and modified Smith predictor were also proposed to improve control performance [94,95]. However, the Smith predictor is model-dependent and can be ineffective if model mismatch is large while applications of PPI are limited to some particular fields only. Furthermore, the Smith predictor is not as generic and widely used as traditional PID. Hence, the proposed techniques are not optimal solutions to address the delay issue in WHNCS. In addition, the model-based approach such as the Smith predictor highly relies on the plant model accuracy. Because of this, continuous plant model update is required. This causes wireless nodes to operate non-stop and their batteries can be drained faster. Therefore, model-based delay and packet dropout approaches are suitable only on the controller side where the gateway always has a constant, infinite wired power supply. It is not appropriate to use them on field devices where the majority of them are battery-powered and have limited power capacity.

2.4.1.1 Wireless Network Delay Components

From a network perspective, there are two main types of delays in WNCS; they are upstream and downstream delays as presented in Figure 2.10. Upstream delay (t_u) is defined as the network delay when sending a signal from a plant to its controller, while downstream delay (t_d) is defined as the network delay when sending a signal from the controller to the plant.

When further analyzing these two delays, the following are the details of delay components: processing delay at controller; transmission delay from controller to gateway; processing delay at gateway; wireless delay from gateway to field wireless device; processing delay at field wireless device; transmission delay from the field wireless device to the plant's devices (actuator or sensor); and processing delay at the plant. From a control viewpoint, although each delay component plays a certain role in affecting control performance of the overall system, the latest works [16,58] in the field pay attention to overall upstream and downstream delays as they provide overall end-to-end information about delay in the WNCS.

2.4.1.2 Delay Causes in Wireless Networks

Some main causes of delay in the WNCS are (i) node joining time [18]; (ii) clock drift [62]; and (iii) individual device internal processes [91]. The join time of a wireless node strongly depends on the number of nodes in the network.

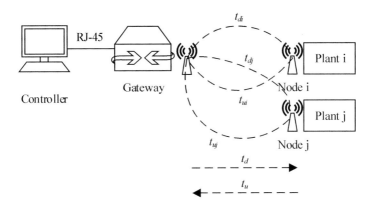

FIGURE 2.10: WNCS main delay components

In general, a network with a single node takes less time for the node to join the network. While a network with multiple nodes takes longer for network formation or for all nodes to join the network. On the other hand, the clock drift is the result of the time difference in network devices' internal clocks; it can be solved by network-wide synchronization [96]. For Linear Technology SmartMesh WirelessHART nodes, the achievable clock accuracy is less than 1 ms [97]. Some important devices in the network are the controller, gateway, plant's sensor, and actuator. Their internal processes contribute mainly to the delay of the network. Wireless delay can be measured accurately by the gateway as it is able to read timestamps of any message in the network. While wired delay can be lumped together with process delay as process dead-time.

2.4.2 Packet Dropout Issues

In WNCS, a common event often investigated when considering control performance is packet dropout [88]. It is defined as an event in which a sent packet fails to reach its destination. A fail-to-reach-destination packet results in packet dropout or packet loss, thus, the destination device could not receive information it needs at the sampling instant. This can cause the process plant to be unstable and uncontrollable if continuous packet dropout occurs. For WHNCS, the packet dropout issue was first reported by Mauro, et al. [61]. Though the digitally proportional derivative (PD) controller of the DC motor was used in the work, its control parameters design was not described. It is possible that packet dropout affects the network's performance [20] especially in burst transmission [90].

2.4.2.1 Packet Dropout Causes

Some common causes of packet dropout are (i) receiver is out of communication range of transmitter; (ii) signal-to-noise ratio is lower than the receiver's sensitivity, thus it is unable to differentiate between a good signal and a noise, resulting in wrongly received packet [98]; and (iii) excessive delay in network communication, which can be due to packet jamming or multiple retransmission mechanism [2]. The first cause can be addressed by rearrangement of physical locations of wireless equipment. The second and third causes are closely related to the wireless device itself and the infrastructure that supports their communications, respectively. Thus, an approach to address these two causes is to use high wireless transmission power and short packet lengths, respectively [88]. However, this approach can drain the wireless module's battery faster, thus shortening its lifetime as well. The following sections will discuss the two common packet dropout models, namely Bernoulli and Markov. They can be used for modeling and simulation of WNCS.

2.4.2.2 Packet Dropout Models

Packet dropout is the main source of instability and poor performance for real-time control systems. Average dropout rate method has been used widely to model the dropout rate of the packet transmission, however, the method could not provide detail on how packet dropouts are distributed [99]. The Bernoulli process shall be used to model the packet dropout process. It has two possible values, 0 and 1, to indicate whether a packet is received or dropped, respectively, at any time. However, the Bernoulli process does not present the dependence of the future status of a packet on the current packet's status. Hence this is a disadvantage of using the Bernoulli process in modeling packet dropout process in wireless communication. According to Hua et al. [99], Markov chain provides more precise representation of stochastic dropout sequences, especially how packet dropouts are distributed. Therefore, the Markov chain model is preferred in the simulation of wireless communication.

In a Markov chain model, the following relationships are set:

$$p_{01} = 1 - p_{00} \tag{2.1}$$
$$p_{11} = 1 - p_{10} \tag{2.2}$$

where p_{00} is the probability that the next packet will be received if the current packet is received, p_{01} is the probability that the next packet will be dropped if the current packet is received, p_{10} is the probability that the next packet will be received if the current packet is dropped, p_{11} is the probability that the next packet will be dropped if the current packet is dropped.

The equations for these probabilities are

$$p_{00} = n_{00} - n_0 \tag{2.3}$$

$$p_{01} = n_{01} - n_0 \tag{2.4}$$

$$p_{10} = n_{10} - n_1 \tag{2.5}$$

$$p_{11} = n_{11} - n_1 \tag{2.6}$$

where n_0, n_1 are the numbers of received and dropped packets, n_{00} is the number of packet pairs that the next packet is received if the current packet is received, n_{01} is the number of packet pairs that the next packet is dropped if the current packet is received, n_{10} is the number of packet pairs that the next packet is received if the current packet is dropped, n_{11} is the number of packet pairs that the next packet is dropped if the current packet is dropped.

For n packets being transmitted from a transmitter to a receiver, the number of packet pairs is $n-1$. The last packet is excluded from the pairs because it is not paired with any other packet. After the Markov chain model, the following equations are formed.

$$n_0 + n_1 = n - 1 \tag{2.7}$$

$$n_{00} + n_{01} = n_0 \tag{2.8}$$

$$n_{10} + n_{11} = n_1 \tag{2.9}$$

When considering the packet series that has been transmitted and received or dropped in the reverse direction, the following property can be found:

$$n_{01} \cong n_{10} \tag{2.10}$$

From (2.4), (2.5), (2.7) and (2.10) it is found that:

$$n_0 \cong \frac{n-1}{1 + \frac{p_{01}}{p_{10}}} \tag{2.11}$$

If p_{01}, p_{10}, n are known, n_0 can be determined using (2.11), n_1 can be determined using (2.7) and p_{00}, p_{11} can be determined using (2.1) and (2.2). For comparison purposes, the Bernoulli process can be used for simulation of the packet dropout process. Thus, the equivalent Bernoulli receiving and dropout probabilities are:

$$p_0 = \frac{n_0}{n} \tag{2.12}$$

$$p_1 = 1 - p_0 \tag{2.13}$$

By using these models, WNCS packet dropout can be simulated. Additionally, since the correlation between the Bernoulli model and the Markov model is established, one can relate these two at the same time for studying packet dropout occurrences and evaluate the models at different simulation conditions. Table 2.7 shows an example of packet dropout arrays having the same size $n = 101$ generated using Markov chain and Bernoulli models [20]. For demonstration purposes, the Markov packet dropout model's receiving probabilities were chosen to be: $p_{01} = 0.6$, $p_{10} = 0.6$. From (2.11), $n_0 = 50$. From (2.12), $p_0 \approx 0.5$. From (2.13), $p_1 \approx 0.5$. Hence, the Bernoulli packet dropout model has dropping probability: $p_1 \approx 0.5$.

TABLE 2.7: Sample generated packet dropout array

Item	Bernoulli Model	Markov Model
Generated Packet Dropout Array	00111001111110101110001 01010011000011111010001 11000011110111011000011 00100111110110010000000 00001	00000000001010011001001 01010101010010011011010 0110011011001101011001 01001010101010110011111 11111

It is important to notice that the occurrence of events in the Bernoulli model is randomized and each event is independent of another. While for the Markov model, the occurrence of one event is linked with its previous or next event, clearly shown at the beginning 0s (continuous received packets) and the trailing 1s (continuous dropped packets) of the generated packet dropout array (see Figure 2.11).

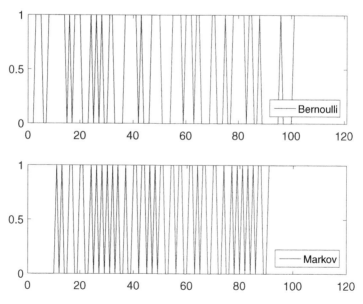

FIGURE 2.11: Packet dropout array generated using Bernoulli and Markov models

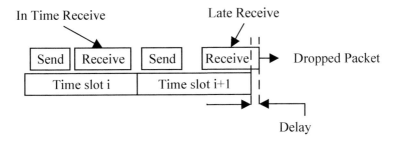

FIGURE 2.12: Delay and packet dropout correlation

2.4.3 Correlation between Delay and Packet Dropout

The correlation between delay and packet dropout is important for assessment of control performance of a WNCS. Ideally, a wireless process plant using technology such as WirelessHART could achieve very high reliability, up to 99.9%. However, the random delay persisting in the network could make the plant unstable due to the late arrival of a control signal, causing the plant to receive not-on-time control packets. A delay signal, if exceeding an acceptable delay threshold, is considered a dropped packet as shown in Figure 2.12. A process plant experiencing this packet dropout should have a mechanism to protect itself from not receiving the control signal, preventing the plant from instability. Furthermore, study of this correlation is highly important especially when wireless technologies rely on TDMA for which their communication is constrained to a pre-scheduled time slot or superframe, a multiple of continuous time slots. This time slot is referred to as the aforementioned acceptable delay threshold.

2.4.4 Energy Efficient Feedback Control Demand

The application of a wireless sensor network (WSN) for process control and monitoring has enabled significant reduction in cables, maintenance and operational expenses. However, the adoption of wireless technologies for a process plant faces a concern on power consumption of the wireless devices and their lifetime durations. A wireless device that requires frequent battery replacement will void the convenience of using wireless devices and will not be preferable in industries. This is because, instead of using cable for power and communication, the process plant needs to monitor the battery more often and the same is applied for maintenance. Despite the wide use of wireless devices for sensor feedback, their use for control has major challenges in terms of battery lifetime. As currently, to achieve desired control performance, the control signal update rate must be 8 to 16 times higher than the feedback update rate [1], which potentially depletes the respective battery 8 to 16 times faster as well. For example, if a sensor feedback wireless node can operate for 4 years and if it is used for control, its operation lifetime potentially reduces to 0.5 or 0.25 year, respectively. One potential approach to address this issue is to power up the wireless device with a constant power supply which in turn requires power cable installation nearby. Another approach is to supply the wireless device with a rechargeable battery and a secondary renewable power supply such as a solar panel. With this, the battery will be charged and its power delivery capacity will be maintained. However, continuous charge and discharge operations can also degrade the battery's lifetime, resulting in degradation of WSN. Therefore, there is an inevitable demand for maintaining and improving the overall performance of the process plant and its WSN with an energy-efficient control algorithm which is simpler, faster and more efficient. One potential candidate is using an EWMA filter, which will be discussed in this research.

2.4.5 Ultra-Low Power Consumption of Wireless Nodes

In a wireless networked control system (or wireless sensor actuator network, WSAN), it is more important to conserve the battery capacity and extend its lifetime than to optimize performance matrices [100–102]. This is because one of the crucial requirements to making an adaptor acceptable by industries is the ability to self-operate with a battery for several years [1]. The works in [21, 103] have discussed battery lifetime and energy consumption of some wireless adaptors. However, the analysis with respect to different operating conditions such as update period of sensor, actuator in WSAN, was not provided. Hou and Zheng [104] studied battery energy consumption on packet-base, which is different from the time-base, e.g., time division multiple access [16] used in job scheduling for WirelessHART devices. On the other hand, the drawback of such an approach is that it does not take into account the power consumption with regard to the pre-scheduled update period of the electronic circuit [105]. Therefore, it is important to develop an approach to estimate battery lifetime of a designed wireless adaptor and optimize its operating condition to meet the critical industrial lifetime requirement. This approach can take into account different update periods of the adaptor, and power consumption for both electronic circuit and communication activities.

Apart from being a native device, each wireless adaptor is required to have the capability of interfacing with the legacy HART-based instrument to ensure the backward compatibility of the protocol. In addition, the wireless transceiver must operate at 2.4 GHz frequency in the ISM band as specified in the WirelessHART standard. The expected lifetime is between four to ten years at typical update period of 8 s [1]. This should be achievable with 3 V or lower of operational voltage of the circuit and 3.6 V of the nominal voltage of the battery.

The three most important components that need to be carefully chosen because they contribute to the majority of power consumption of the nonnative WirelessHART adaptor (see Figure 2.13) are (i) radio transceiver, (ii) microcontroller (MCU), and (iii) HART modem. To theoretically achieve the aforementioned requirement, the radio transceiver shall be selected as a single-chip, ultra-low power consumption nRF24L01+ IC operating at the ISM band with maximum data rate of 2 Mbps [106]. The MCU shall be STM32L051x6 [107]. It supports ultra-low power consumption applications suitable for the wireless sensor network (WSN). The MCU offers dynamic power supply with three voltage levels: 1.8 V, 3.0 V, and 3.6 V. Additionally, it has an internal voltage regulator that regulates the supplied voltage to MCU dynamically. This feature provides a range of options for electronic design engineers to consider and select an operational mode and a voltage level that best fit their specified applications. The HART modem for interconnecting a wired device with the wireless adaptor can be DS8500 [108] which is certified by FieldComm Group, the merger of Fieldbus Foundation and the HART Communication Foundation. The power supply for the adaptor shall be a pack of five parallel heavy duty, low self-discharge, wide operating temperature range and prolonged lifetime LS 33600 batteries from Saft [109]. It is made of lithium-thionyl chloride which is suitable for industrial applications [1]. The nominal voltage and capacity of each battery are 3.6 V and 17.0 Ah, respectively. A selected MCU's operating voltage of 3.0 V can ensure there is a suitable voltage difference between the power supply and the circuit for reliable operation of the adaptor [56, 110].

When this sample WirelessHART adaptor is used, the estimated battery lifetime [56] is presented in Figure 2.14. As seen from the figure, the most suitable operational frequency of the MCU is 4 MHz. This yields the longest lifetime (of 4 years) for the adaptor given the selected battery. This lifetime can be extended by choosing a battery with higher capacity, however, it should be noted that the battery generally degrades over time and with the number of charge or discharge cycles. When the update period is longer or the adaptor is used less frequently for wireless communication, the battery lifetime increases. Eventually, it will reach a saturated state at which there is no significant improvement in the battery's lifetime.

2.5 Summary

In summary, this chapter has covered the introduction of industrial networked control systems with main focus on wireless networked control systems. It further introduces WirelessHART protocol, the leading technology for enabling wireless communication and control of industrial process plants. An introduction to WirelessHART simulators is presented for use as an important tool

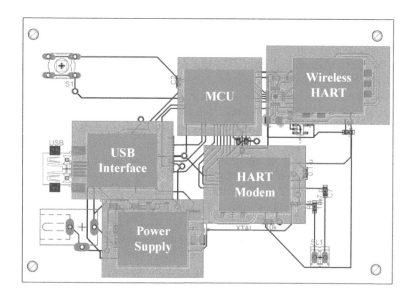

FIGURE 2.13: Sample WirelessHART adaptor circuit

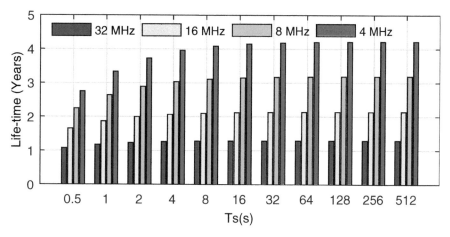

FIGURE 2.14: Estimated battery lifetime of sample WirelessHART adaptor circuit

in learning and researching the technology in both laboratory and industrial scales. The chapter then discusses several challenges for applying the technology in industrial process plants. The two most common issues are delay and packet dropout which can severely affect control performance of the plants. The close correlation between the two issues is highlighted to provide better understanding of this issue. Furthermore, the discussion on the needs for energy efficient feedback control together with ultra-low power consumption of the wireless adaptor is provided. In the next chapter, a more detailed look into the uncertainties in industrial wireless mesh networks will be presented. These uncertainties contribute to the mentioned delay and packet dropout happening in the industrial network. Thus, reducing these uncertainties through the means of wireless communication technology development helps to maintain good quality of the wireless network.

3

Uncertainties in Industrial Wireless Mesh Networks

3.1 Introduction

This chapter discusses the uncertainties as well as time varying behaviors of the wireless channels in a mesh network environment. The chapter starts with a discussion of common environment noises in the networked control system. Then it elaborates on the changing temperatures of the surrounding harsh environment which significantly results in path fading and overall wireless signal quality. Furthermore, problems related to channel fading will be introduced. Signal-to-noise ratio (commonly known as SNR ratio) is also one important factor that determines the quality of a wireless network. Thus, it will be discussed and analyzed in the context of wireless networked control systems. Several quality-of-service parameters and how to measure the wireless link performance will also be outlined. The impact of wireless communication in terms of data rate, packet losses, probability of outage, jitter, etc.. on the closed loop control system performance and control design are then investigated.

3.2 Environment Noises

Environment noise is an unavoidable parameter in any form of communication. It always exists and has several forms such as white noise, Gaussian noise, path loss, temperature surface, construction object, metal frames, etc. These can cause the effective communication range of a wireless device to decrease from as long as 228 m down to about 30 m [55]. By reducing the effect of the environment noise, it is possible to improve the overall wireless communication stability as the signals are less noise-affected. For example, in order to reduce the path loss effect, in practice, when there's no obstruction, the wireless device should be placed 2 m above the ground [55].

3.3   Surrounding Temperature Surfaces

One common source that creates significant noise to wireless instruments is the thermal temperature surfaces nearby industrial instruments [111]. The fast changing thermal surface is considered one of the main causes to fast air turbulence, resulting in major path loss in wireless communication [112]. To avoid this issue, the wireless instrument should be placed away from the high temperature sources such as boilers, heat exchangers, etc. [55]. This helps to improve overall wireless communication and reduce path loss. In addition, the wireless communication will be more stable as interference is reduced.

3.4   Path Fading

In the real world, various interferers do exist in process plants such as multi-path fading, noise from machinery equipment, non-line of sight communication, object blocking. Interference is very common especially for a network with open Industrial Scientific and Medical (ISM) band (2.4 GHz). This is because the band is license-free and is usable by all types of devices such as mobile phones, radio devices, Wi-Fi and Bluetooth devices, etc. Therefore, due to signal attenuation, obstacles in radio path, multiple signal reflections, and communication propagation delays, fading and phase shift, system delays (e.g., sensor/controller delay, controller/actuator delay), path loss occurs. This can severely affect quality of a wireless networked control system, particularly in an industrial environment. In addition, packets dropping or being discarded will create issues with the network as well [46]. It should be noted that when wireless local area network (WLAN) is close to the WNCS, it will completely destroy the low power wireless signal [51]. One of the typical parameters that determine the path fading effect is the received signal strength indicator (RSSI). Its relationship with distance depends on many unpredictable parameters. For example, in IEEE 802.11 network, a small change in direction or position results in significant changes in RSSI. The existence of moving objects in the surrounding environment also introduces changes in RSSI even without location change. Distance is obviously a parameter that affects RSSI since it causes attenuation (or path loss) in transmitting the wireless signals.

Another factor is multi-path (fast fading), shadowing (slow fading). Through averaging the value over a certain time interval, it is possible to eliminate fast fading. Though the path fading coefficient n can vary, it is suitable to assume that value to be constant at short intervals of time. α depends only on radiation characteristics of the device while n varies and depends on environment and can change abruptly from time to time. Thus, measurement of alpha can be done with several measurements and it can be considered as a constant value for different surrounding environments.

In order to overcome the narrow band interference such as multi-path fading, FHSS is one of the effective mechanisms. It is implemented as part of the core feature of wireless mesh technology such as WirelessHART. A simulation with the formula can show that path loss increases and received signals power decreases when distance between transmitter and receiver increases and vice-versa. In indoor environment, path loss is lower than that in outdoor environment. Therefore, packet dropout and path loss should be taken into consideration of designing a wireless networked control system to improves the systems reliability. This also addresses challenges for WNCS as wireless data transmission has potential to lose the connection, when feedback is lost or when control action does not reach actuator due to the loss of connection, control action shall not be taken.

3.5 Signal-to-Noise Ratio

Given any communication system, one of the key factors that determines the quality of signal transmitted is signal-to-noise (SNR) ratio [113,114]. Depending on the type of receiver, their sensitivities, the quality of received signals can be determined through its SNR parameter. This is the representation of how strong a received signal is compared to the environment noise's power. The higher the ratio, the easier the receiver receives the signals or alternatively, the more sensitive to the signals the receiver is. As noise always persists in the environment, in an industrial network, due to the harsh nature of the surrounding, the SNR ratio can significantly differ from one wireless node to another. This is because often the nodes are scattered from each other. In order to improve the SNR, location of the wireless nodes (sensors or actuators) should be selected to avoid thermal noise, noises from rotating and other wireless equipment, and avoiding metal frames or casings, as these contribute significantly to the generated interferences [54,55].

3.6 Overall Network Traffic

Given any network, the first thing that should be done is the network setup, physically and programmatically. The latter means engineers need to perform network setup and configuration using the settings provided by the network manager, often integrated into the network access point or router. When considering the network as a whole, network traffic can serve as an indicator of network performance. Depending on the need of each individual application in the network, traffic to and from a node in the network is configured accordingly. Over-speeding the network traffic results in several drawbacks as follows:

- Local contention and congestion at the nearest gateway or router [35];

- Imbalanced network load [100, 115];

- Service degradation due to local contention and congestion [12];

- Potentially high rate of transmission failure or packet dropout [2, 12].

All these factors can cause severe degradation of overall network performance. Thus, network traffic should be configured properly in the wireless networked control system. In general, more speed and bandwidth should be allocated for control paths as it is critical while less can be assigned to feedback paths. In addition, prioritized network traffic shall be implemented to provide queue-with-priority support for processing more critical messages first. The higher priority shall be applied to control signals, or signals that trigger events.

3.7 Summary

Overall, this chapter has discussed several factors that contribute to the uncertainties of industrial wireless networked control systems and some approaches to mitigate the uncertainties. However, the suggestions are mainly through physical changes or setup, or appropriate selection of locations. In the following chapter, a novel technique for handling noises using a modified exponentially weighted moving average filter is presented. The technique relies mainly on the way the system processes the received control signals rather than by changing its physical properties. Thus, it can be applied widely in industrial wireless networked control systems and is less dependent on the environment setup. As long as the signals are available to the receiver, it is possible to improve overall network control performance.

4

Delay and Packet Dropout Measurement in Wireless Networked Control Systems

4.1 Introduction

In an industrial environment, as discussed in the previous chapter, wireless networked control systems experience delay in communication and to some extent, it can cause packet dropout. This in turn results in control system performance degradation. In order to tackle these issues, it is essential to realize realistically the amount of delay and packet dropout persisting in the network. Therefore, this chapter focuses on the measurement aspect of the delay and packet dropout of industrial wireless networked control systems.

4.2 Wireless Network Delay Measurement Experiment

In this section, two delay measurement approaches are presented. They are direct end-to-end delay measurement using gateway and wireless node, and indirect one-end delay measurement using only gateway. End-to-end measurement provides benefits for realizing the exact delay between any source-destination device pair. While one-end measurement is most suitable for controllers to optimize the control parameters based on the measured average delay in the network. Thus, control performance of the system can be improved if the delay is compensated by the controller prior to sending a control signal to the plant.

4.2.1 Experiment Setup

Figure 4.1 presents the experimental setup for measurement of network induced delays. As seen, the setup consists of four components: controller, plant, gateway and wireless node. The controller is connected to the WirelessHART gateway through LAN cable. The gateway communicates with the node using WirelessHART standard. The node is attached to a plant which can be either a computer, or a microcomputer such as Raspberry Pi 2 Model B, or a micro-controller such as Arduino Mega 2560. The communication between the node and the plant is serial. Both gateway and nodes are manufactured by Linear Technology [116]. This setup can be used for both end-to-end and one-end delay measurements. A detailed comparison of the two approaches can be found in Section 4.2.5.

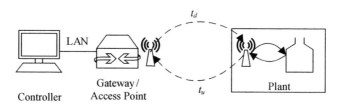

FIGURE 4.1: Network induced delay measurement experiment setup

4.2.2 Supported Communication Modes

The Linear SmartMesh software development kit (SDK) used to perform round trip delay measurement supports different modes of operation as shown in Table 4.1. The node has four modes to transfer data to the gateway namely Publish, Event, Maintenance, and Block Transfer. On the other hand, the gateway only has Maintenance mode. The Publish mode is associated with sending sensor data from node to gateway. The Event mode is dedicated for creating an event in the network. The Maintenance mode is used for setting the network and broadcast message to all nodes and Block Transfer mode allows the node to function as a relay, an intermediate device that routes data from one node to another. Typically, the node has Low, Medium, and High communication priorities while the gateway has only Low and High priorities.

TABLE 4.1: Supported WirelessHART communication modes

Mode	Node Priority			Gateway Priority	
Publish	Low	Medium	High	-	-
Event	Low	Medium	High	-	-
Maintenance	Low	Medium	High	Low	High
Block Transfer	Low	Medium	High	-	-

4.2.3   End-to-end Delay Measurement Experiment

In order to perform delay measurement of a wireless plant, it is assumed that the plant is connected to a wireless node which acts as a communication point for both transmitting and receiving signals to and from the gateway. The network induced upstream and downstream delays will be measured by continuously sending data from the node to the gateway and from the gateway to the node, respectively. Once the destination receives the sent data, it will automatically create a timestamp indicating the moment the data are received. Thus, by calculating the difference between the two consecutive timestamps, the network induced delay is determined.

When the Linear SmartMesh SDK program is run at the controller, it will measure the upstream delay. When it is run at the plant, it will measure the downstream delay. Upstream delay is the delay in communication from the node to the gateway, while downstream delay is the delay in communication from the gateway to the node. By analyzing the logged data, the network induced delays at different modes of operation can be investigated.

In order to start the experiment, firstly, the controller turns on the gateway allowing the wireless node to join its network. Secondly, the application programming interface (API) explorer program in the SDK is opened to automatically record data sent from the node. Then, the plant turns on the node and connects it to the network formed by the gateway. Through the node, the plant then continuously sends two-byte data to the gateway. The two-byte payload is sufficient for transferring a single sensor parameter over the network as its value ranges from 0 to 65535. Each communication mode is performed 101 times to produce 100 upstream delay measurements. To measure the downstream delays from the gateway to the node, a similar approach is followed.

4.2.4   One-end Delay Measurement Experiment

Similar to the end-to-end measurement approach, one-end measurement uses the same hardware setup and software as shown in Figure 4.1. However, one-end measurement can only be done at the gateway as it has transmission information of each and every packet routing in the network. The measured delay is the delay calculated by the gateway based on timestamps of sent and received messages at both upstream and downstream communications. This delay information is useful for optimization of the controller or network as optimization using the delay instant after each transmission (end-to-end delay) is not meaningful due to the time variant nature of the delay. Once the network is established, the delay can be measured by calling a function to get latency from the gateway to a specific device in the network with its unique MAC address. Both upstream and downstream delays can be obtained using the function.

4.2.5 Comparison of End-to-end and One-end Delay Measurement Experiments

Further elaboration on the similarities and differences between end-to-end and one-end delays is described as follows. Both delay types represent overall wireless network delay. The delay obtained from each experiment can be used to calculate overall round trip delay in the network. However, the two delays are different from each other at various parameters such as location, type of delay, requirements to obtain delay, delay characteristics and applications for utilizing the delay as shown in Table 4.2.

TABLE 4.2: End-to-end vs. one-end delay measurement experiments

Comparison	End-to-end Delay	One-end Delay
Similarity	Can be used to calculate round trip delay, Represent WirelessHART network delay	
Difference		
Availability Location	At gateway for upstream delay, at node for downstream delay	At gateway for both upstream and downstream delays
Type of Delay	Immediate delay instant (after a transmission)	Average delay instant (an average value of all captured delays over a finite time horizon preset by the gateway)
Requirements	Establishment of round trip communication between two devices, Continuous delay sampling	Establishment of round trip communication between two devices
Characteristic	Time-varying with possibility of abrupt change (e.g., a jump from 2 s to 5 s)	Time-varying without abrupt change
Application	Delay compensation at each end device, prior to sending out signal	Delay compensation at controller side

End-to-end delay is available at a respective end device while one-end delay is only available at the gateway. The end-to-end delay is an instant delay which means it varies over time and there is always a difference between any two consecutive delay instances. Thus, it is possible to have an abrupt delay change in the end-to-end delay, while one-end delay varies over time as well. However, the change in delay is introduced more smoothly since the effect of a sudden change in a delay instance is averagely distributed. Furthermore, end-to-end delay measurement requires a continuous sampling process and a round trip communication between any two devices, while one-end delay measurement only needs to establish the connection between the two devices and the measurement is available at gateway, often collocated with controller. In addition, the measurement done by the end-to-end approach does not represent well the next measured delay because it becomes outdated right after the measurement is performed, while one-end measurement reflects the averaging of the recent delays, thus it also represents the trend of delay for the upcoming measurements. Therefore end-to-end measurement is more suitable for assessing the communication between two devices for a particular situation, while one-end measurement is more suitable for use in controlling the plant.

4.2.6 Effect of Variable Payload on Network Induced Delay Experiment

It can be recalled from Section 2.4.1.2 that the identified causes of delays in WNCS are connection establishment, node joining, network maintenance, moving node due to equipment vibration [18]; message payload construction and delay [117,118], clock drift [62]; and internal device delay [91]. In this section, an experiment is described to study the correlation between the wireless control message delay and its variable payload length.

The experiment involves three main components: a controller (a computer with controller application), a WirelessHART gateway, and a WirelessHART node as shown in Figure 4.2.

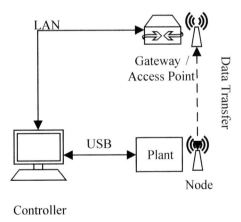

FIGURE 4.2: Variable payload effect on network induced delays experiment setup

The controller is connected to the gateway using a LAN cable. The node is assumed to be connected to a plant. In this case, it is connected to the same computer as the controller using a USB cable. The communication between the gateway and the node follows WirelessHART standard [10] since both devices are WirelessHART certified. The selected communication domain is Public which is mainly used for broadcasting messages from the gateway to all nodes in the network. Since the guarantee for receiving the transmitted message is not considered in this book, Best Effort communication is configured. In general, if the sent message has high priority, it will be scheduled to be sent first, thus, minimum transmission delay can be achieved. All packets generated and transmitted are marked with High priority. Subject to randomly generated payload, the network delays are recorded and analyzed to identify the relationship between the payload length and the induced delays.

4.2.7   Round trip Delay Calculation

Round trip delay is usually used for control study of simplified WNCS. This is termed as lumped delay since it is the summation of both upstream and downstream delays. The formula for calculating this delay is shown below.

$$t_{rd} = t_u + t_d \tag{4.1}$$

where t_{rd} is round trip delay, t_u, t_d are upstream and downstream delays.

4.3   Wired Link Contention Experiment

Network reliability depends on the reliability of the wired and wireless communications. If contention occurs, network delay will increase, causing packet dropout because a scheduled control system can only accommodate certain delay tolerance. Since WirelessHART network is reliable, an experiment to study wired link contention between wireless node and computer running MATLAB®15 is presented here. The communication between the computer and the wireless node represents well that between a controller and a gateway, or a wireless node and actuator. The experiment procedure is described in Figure 4.3. Through the experiment, the contention or effect of different continuous request delays on packet dropout rate over a wired link between the devices is studied. The sent and received packets are High-Level Data Link Control (HDLC) frames because they are used for wired communication between the devices.

HDLC frame has a flag sequence (i.e., 0x 7E) to indicate the beginning and the end of the frame, a frame payload and a 2-byte frame check sequence (FCS) to ensure the frame's integrity [117]. To prevent the flag sequence from

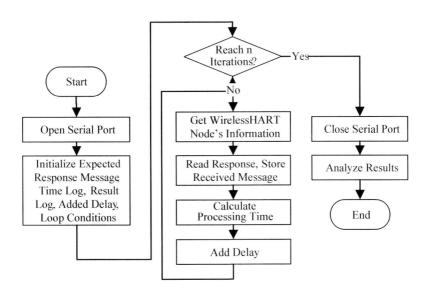

FIGURE 4.3: Wired link contention experiment procedures

appearing elsewhere in the frame payload, byte-stuffing is used after the calculation of FCS. If the flag sequence is 0x7E, it is replaced by 0x7D5E. If the control escape is 0x7D, it is replaced by 0x7D5D. In WirelessHART, 16-b ITU-T polynomial is used to calculate the FCS [53]. The calculation can be done using techniques such as digital logic, modulo-2 arithmetic, polynomial and inverse polynomial [119]. In this experiment, FCS calculation using inverse polynomial is performed.

For starting the experiment, the typical serial port with configurations of 9600 bauds, 8 data bits, no flow control, no parity, continuous asynchronous read mode, 1 stop bit and newline terminator is opened for the controller to communicate with WirelessHART node. The validation response message with content 7E 02 11 01 00 0C 00 00 17 0D 00 00 1A AE FC 02 25 02 02 01 00 07 3A 57 7E [117] and necessary simulation variables are initialized. The loop for sending a message to and receiving a message from the node starts with stopping condition of 1,000 iterations. The program sends a command to get the node's information which is 7E 02 01 00 0C 18 55 7E. It waits for a response from the node, and the execution time is captured. After this step, the program intentionally inserts an added delay. This is the delay between any two consecutive send requests. The operation is continued until the loop is terminated. Once the loop ends, the serial port is closed and all captured data are analyzed. The operation is applied for different added delays ranging from 0 s to 0.50 s with a step change of 0.05 s. For a fast sampling control system with a sampling period of multiples of 0.01 s (10 ms), the above delay range is suitable for the experiment. It should be noted that the added delay is not accounted when estimating execution time as it is after the Log Execution Time action.

Often, the wired link delay due to the controller is not the same as the added delay in the experiment, hence, an interpolation process is used to estimate packet dropout with the knowledge of wired link delay. When known boundary packet receiving rates and added delays are p_{r1}, p_{r2} and t_{d1}, t_{d2}, if wired link delay is t_d, the estimated packet receiving rate is described in (4.2). It will be used in Section 8.3.2 to estimate the packet receiving rate.

$$p_r = p_{r2} + \frac{p_{r1} - p_{r2}}{t_{d1} - t_{d2}}(t_d - t_{d2})$$ (4.2)

The relationship between packet dropout rate and packet receiving rate is presented in (4.3).

$$p_d = 1 - p_r$$ (4.3)

where p_d is packet dropout rate.

With the knowledge of packet receiving rate, the packet dropout rate can be calculated using (4.3).

4.4 Summary

In summary, this chapter has discussed the two main issues associated with wireless networked control systems, namely delay and packet dropout. These contribute to the significant degradation of network control performance, thus their effects on the system should be minimized. The following chapter will further discuss several techniques that are available for this purpose.

5

Delay and Packet Dropout Compensation Techniques

5.1 Introduction

As mentioned earlier in this book, the two main issues with any wireless networked control system are delay and packet dropout. Both issues are inter-related, and they can affect networked control performance severely [2, 120]. In order to counter their negative effect on the control performance of a networked control system, the following sections will discuss potential approaches to address them using techniques such as the Smith predictor and the Kalman filter, etc.

5.2 Delay Compensation Techniques

Several related works have attempted to address the delay issue of WNCS using approaches such as the Smith predictor [59], the Kalman filter [121–124], the EWMA filter [125], and the EWMA controller [126–128], PID-PLUS [1, 129]. Key drawbacks of the control schemes using the Smith predictor and the Kalman filter are complexity and prolonged processing time, especially with low-processing-capability, energy-constrained microcontrollers. This type of microcontroller is used in field instruments for preserving battery lifetime over a long period. On the other hand, the EWMA control approach is mainly applied in semiconductor industries instead of process industries. Thus this book will investigate the suitability of EWMA filter for delay and packet dropout compensation in industrial process plants. The following section presents a more detailed review of these techniques, their advantages and disadvantages.

5.2.1 Smith Predictor

Without considering disturbance, the simplified Smith predictor controller developed in 1957 is represented in Figure 5.1 [87]. In the figure, $R(s)$ is the system's desired setpoint; $Y(s)$ is the process output; $C(s)$ is the controller; $P(s)$ is the process plant transfer function; $\tilde{P}(s)$ is the model of the process plant; θ is the process dead-time.

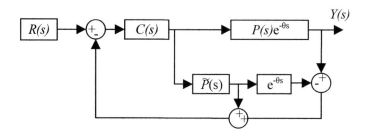

FIGURE 5.1: Simplified representation of Smith predictor controller

Without model mismatch $\tilde{P}(s) = P(s)$, the closed-loop transfer function is $\frac{C(s)P(s)e^{-\theta s}}{1+C(s)P(s)}$ while that of conventional feedback control is $\frac{C(s)P(s)e^{-\theta s}}{1+C(s)P(s)e^{-\theta s}}$. It is seen that a Smith predictor controller helps to eliminate the delay component at denominator of the transfer function as compared to the traditional feedback control. Even though a Smith predictor can provide superior performance improvement over the conventional feedback control system, it is only from a theoretical aspect. Furthermore, since a Smith predictor controller is a model-based approach, the performance improvement relies heavily on the accuracy of the model. An inaccurate process model can result in significant control performance degradation, up to an instability condition [59]. Hence, this is the key disadvantage of Smith predictor and any other model-based filter such as a Kalman filter [121–124].

5.2.2 Internal Model Controller

Several internal model controller (IMC) methods have been employed to solve a diverse range of problems from delays, load change [130], uncertainties [131] to disturbances [132–134]. However, there is no reported case of its application for WNCS, particularly for the WirelessHART system. The basic IMC structure is presented in Figure 5.2. In the figure, different from the traditional controller, IMC controller has a plant model $\tilde{P}(s)$, placed in parallel with the real process plant $P(s)$. This is to predict and calculate responses of the virtual plant subjected to the same control conditions as the real plant.

The difference between the output of the real plant and the virtual plant is fed back to compare with the setpoint. This helps to improve the overall control performance of the system, as long as the process plant model is rel-

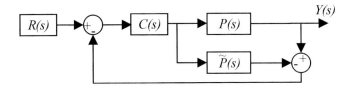

FIGURE 5.2: Basic IMC structure

atively close to the real process plant. In general, there are two main steps to design suitable parameters for an IMC controller [134]. The first step is to factorize the process model $\tilde{P}(s)$ into two components: one contains time delays and right-half plane zeros with steady-state gain of one $\tilde{P}_+(s)$ and the other one $\tilde{P}_-(s)$ is the remainder of the model. The second step is to calculate controller transfer function $C(s) = \frac{f(s)}{\tilde{P}_+(s)}$ where $f(s) = \frac{1}{(\tau_c s+1)^n}$ is low-pass filter with steady-state of one. In the filter's transfer function formula, τ_c is the desired closed-loop time constant, and n is the filter order. Usually n is an integer value that is carefully selected to cancel the order of $\tilde{P}_-(s)$. Since the dynamics of a control system, in reality, is time-varying, its transfer function will not be constant. Therefore the selection of a suitable value for n is a challenging task.

From a WHNCS perspective, with the help of cyclic redundant check, a wireless message delivered to a process plant will be checked to ensure its integrity before the control or feedback signal is used. Therefore, a wrongly received message at the plant's input will be discarded. Hence, the system delay [16, 58, 83] and packet dropout [61] are the key issues that should be addressed. Despite the effectiveness in disturbance rejection, the IMC controller faces similar issues as the Smith predictor since both are model-based approaches. Depending on the accuracy of the model used in the system, the performance of the techniques will be affected.

5.2.3 Fuzzy Interference

Fuzzy-PID controllers have been well studied in recent research with focuses on the gain tuning mechanism, controller parameter tuning, comparison with other controllers such as PID, and model predictive controllers (MPCs) using different performance measure indices. An optimal fuzzy PID controller in comparison with a PID controller is studied in [135]. The controllers were tuned by using the integral of time multiplied absolute error (ITAE) and squared controller output for addressing randomly time-varying delay issue in NCS. The tuning mechanisms used in this research are genetic algorithm (GA) and particle swarm optimization (PSO). The challenge for optimization using these techniques is in the selection of appropriate population size. This is because when the population size grows it may result in better performance, however, at a higher cost of computation. Furthermore, particularly in the work, for PSO approach, only star and ring network topologies were considered. They are only subsets of the complex mesh network topology offered by WirelessHART technology.

A fuzzy logic system was designed for adjusting virtual resistances of droop controllers in a low voltage DC micro-grid [136]. The system guarantees balancing of energy stored at distributed batteries based on their state of charges. However, detailed fuzzy logic control design was not presented. Takagi - Sugeno - Kang (TSK) neuron-fuzzy network was used to obtain the process model's key parameters in [137]. Additionally, a fuzzy adjustor was developed to form part of the optimal setting controller which initializes a set of initial setpoints for the overall process to achieve desired operational requirements. It was used to generate a new set of setpoints based on ongoing process states when the particle sizes of soft sensor modules (part of optimal setting controller) drift away from the required ranges. Despite the common use of fuzzy logic controllers in recent control systems of nonlinear complex processes, the fuzzy logic designs are mainly based on experts' knowledge of the systems and it is difficult to derive or prove the designs using mathematical expressions [137].

5.2.4   EWMA Controller and Filter

The application of EWMA in the semiconductor industry has been broadened to two main areas: using EWMA as a filter [125, 138, 139] and as a controller [126–128, 140, 141]. The EWMA controller is being used widely in the semiconductor industry for run-to-run (R2R) control due to its effectiveness and simplicity [142]. Additionally, for the process with both advanced process control (APC) and R2R control, Chen-Fu, et al. [143] developed a dynamically adjusted PI (DAPI) R2R controller to outperform the traditional controller used in a semiconductor company. Usually, the process gain (EWMA-G) is used as off-line estimate while the intercept term (EWMA-I) is updated for each run. As related to this, Jin [127] carried out studies of stability and sensitivity of EWMA-G controller in comparison with the more well-studied EWMA-I controller. Since foundries need to spend additional operating cost for hiring experienced operators to initialize independent bias values for each new thread, hybrid initialization using both threaded and non-threaded controls was proposed by Harirchi, et al. [141].

Often, EWMA controllers are designed with a fixed discount factor. A small discount factor can ensure long-term stability of the system in regular conditions, however, it prolongs the time taken to reach the system's desired setpoint. This does not benefit the small batch process much. Therefore, to shorten settling time, dynamically adjusted discount factor was proposed to accommodate the customization needs [142]. At first, a discount factor is chosen to be large so that it can reach the near-setpoint point faster; then after that, the discount factor is reduced as the system's output is reaching the setpoint. Also while working on the variable EWMA controller, Jou et al. [128] designed a controller to address semiconductor process disturbance due to step changes and linear drift. It is proven that the controller helps to reduce total mean square error of process output even when process parameters are not well estimated. The main disadvantage of the group and product (G&P) EWMA controller is that it requires iterative controller parameter fine tuning which results in additional delay when issuing control command at the controller [144]. Since the intelligent R2R double EWMA (d-EWMA) control strategy proposed by Chyi-Tsong and Yao-Chen [126] requires the use of a genetic algorithm to find optimal discount factor, the approach introduces additional delay in issuing control command from the controller. It should be noted that any additional delay is always an unwanted factor in any process control system. From the filtering perspective, even though the EWMA filter has been used for process plant power load forecasting, the filter parameters design has not been described [125].

Although an EWMA controller with variant discount factor is promising for a control system, it faces a challenging concern for implementation in oil and gas plants in which the majority use traditional controllers such as PID. The alternative is to preserve this kind of controller while using EWMA as a filter. This approach has several advantages such as cost saving, and ease of maintenance and operation since replacement of the traditional controllers is not required. Hence, using the EWMA filter as an essential part of the process plant with its main purpose for filtering signals is proposed in this book. It will be used as an essential component to effectively address packet dropout issue in the network.

5.2.5   PID-PLUS

Process instrument manufacturers approach wireless control using the traditional controller from various ways. In wired control, many products are designed to create a reset component using a positive feedback and feed-forward network. The filter's time constant reflects reset time and is defined in seconds per repeat. From a practical approach, to achieve good control performance when the measurements are not updated periodically, the controller is modified and known as PID-PLUS to address the reset characteristic in wireless control. In the case of both integral and derivative mode, the PID-PLUS diagram is shown in Figure 5.3 [1, 129].

PID-PLUS controller improves overall performance of a wireless measurement process plant while preserving the widely used PID controller [1]. However, the dependence on sensor measurements and high update rate for the actuator, e.g., 16 times faster than the measurement update rate when using

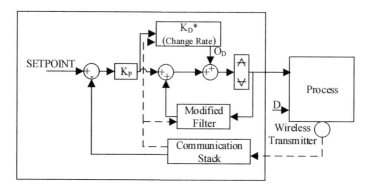

FIGURE 5.3: PID-PLUS controller with integral and derivative mode

the PID-PLUS control scheme, is the main issue impeding its usage in process plants with wireless control application. This is because the higher the update rate, the faster the field device's battery depletes, and thus the shorter its lifetime. Therefore, the wireless actuator's battery needs to be monitored and replaced often, which is inconvenient for plant operators. Alternatively, the device can be powered by a constant wired power line, or it is equipped with a secondary renewable power source such as a solar panel. Even with this option, not all places at the plant are suitable for placement of a solar panel, especially the difficult-to-access spots where a battery-powered device has significant advantage.

5.2.6   Kalman Filter

Developed in 1960, the Kalman filter sparked huge interest and played an important role in many engineering fields, and since the 1970s [145], it has proved useful for widespread applications. The applications range from economics for sales forecasting with time-varying parameters [146], image processing to control applications such as sensorless control of nonlinear systems [147], sensorless [148–150] and sensor [124] speed control of permanent magnet synchronous motors, sensorless control of induction motors [151], sensorless control of AC drives [152], nonlinear high-speed object's state estimation [153]. For a process with noise, the Kalman filter is the optimal state estimator [147].

In nonlinear systems, extended Kalman filters based on linearization of the system's using first-order Taylor expansion can be used [147]. A derivative-free Kalman filter approach [147] is suitable for state estimation in sensorless control of nonlinear systems, i.e., DC motor control. It estimates the state vector of the system without derivatives and Jacobian calculations. In a sliding mode control of wheeled mobile robots, an adaptive unscented Kalman filter was used to track the slip ratio of the robots [154]. It estimated the vehicle's longitudinal velocity and wheel angular velocity subject to the system's parameter variation and disturbances. In a thermal power plant, Kalman filter was used to estimate temperature of the furnace based on plant inputs, outputs and plant knowledge since direct measurement of temperature was absent and an exact model of the plant was too complicated and inaccurate [155].

In a vision-based position control application with presence of measurement noise and disturbance, a Kalman filter was used to provide better state estimation of an object's position and to control it more accurately [156]. Other applications of the Kalman filter include flow rate estimation in an automatic pouring robot [157], state estimation for voltage control in distribution network [158], real-time power system state estimation [159], tracking surface vessels [160], predictive maintenance for epitaxy processes [161], parallel dynamic state estimation [162].

The network path delay metric is important for network operators to assess, plan and diagnose faults [163]. It is one of the most important parameters to measure reliability and timeliness of real-time communication given a delay threshold [91]. Monitoring of all network paths in a large network is a challenging task and it wastes resources. The reason is that the delay and packet dropout path matrices generally grow as the square of the nodes in the network [163]. Hence network performance matrices should be obtained using a statistical approach for a subset of nodes. Rajawat, Dall'Anese, and Giannakis [163] introduced a dynamic approach to estimate network delays using measurements of only a few paths using a spatiotemporal Kalman filter, kriged Kalman filter (KKF). The delay estimation utilizes both historical data as well as the network's topology. The online optimal path selection helps to minimize per-time-slot prediction error. The linear predictor is optimal and the experimental results with real-world data sets were promising. For constant delay compensation of tilt sensor, the fusion sensory system including the tilt sensor and a gyroscope was post-filtered by using a modified Kalman filter [164]. The system model is a discrete linear time-invariant (LTI) with zero-delay measurement. In the work, the Kalman filter has been modified to introduce a compensation term to the standard Kalman filter for measurement delay compensation. However the solution is applied for a single plant instead of a mesh plant network. To some extent, non-continuous delay represents potential packet loss. For a continuous LTI system with time-varying delay, suboptimal Kalman filter based on the minimum estimation variance was used for delay compensation [85]. The system was applied for both bounded delay and non-continuous delay. In the case of the continuous delay, the system offers better control performance as compared to the system that introduces artificial delay to obtain the uniformly distributed measured delays. Additionally, the measured delays in the research are obtained through the timestamp on the received packets.

It is seen that the Kalman filter is promising in performing state estimation of both nonlinear and linear-systems time-varying systems, thus, it can be used for delay estimation in the wireless plant. However, the main issue with the Kalman filter is its high cost of computation due to its iterative nature, which is not suitable for microcontroller-based field instruments.

5.3   Packet Dropout Compensation Techniques

Generally there are two approaches to address packet dropout in WNCS. The first approach is to configure a network (or network devices) to handle the delay through retransmission of an unsent or fail-to-reach packet [91]. However, usually a plant has devices from different vendors, and each has different device designs and procedures to configure. Thus, it is difficult and time-consuming for plant operators to modify each device to include this mechanism.

The second approach is to use a filter for processing of WNCS signals upon receipt. An example is to use the Kalman filter to address multiple packet dropouts, random sensor delays and missing measurements [121, 165, 166]. In WNCS, measurement signals are more prone to transmission delays and thus, packet dropouts. Hence, estimation of measurements is essential to provide adequate information to the control system. An application of the Kalman filter in a plant for state estimation in polymerization processes is described in [167]. Similar to the delay compensation approach using the Kalman filter, it is desired to know plant and controller models for state estimation of plant/controller when control/feedback packet is dropped. The computation with this approach is complex and not suitable for low processing of a wireless microcontroller used in plant actuators. Therefore, an EWMA filter is a potential candidate for addressing these issues since it is computationally efficient, model-independent, and simple to implement.

5.4   Summary

The Smith predictor is well-known for its suitability in addressing time-delayed control systems [87, 168–170] however, one needs to have an exact white box model of the plant, which is ultimately difficult to obtain especially for modern plants that consist of thousands of pieces of equipment. Similarly, the Kalman filter [121, 153, 171] can be used to estimate the content of lost packets. Again, it requires a plant model or packet dropout with respect to the network condition. Hence, this limits the applications of the Kalman filter in addressing the packet dropout issue. Replacing a widely used controller in a process plant with a totally new controller like EWMA is an alternative. This introduces additional cost for replacing existing components in the process plant and most likely will not be a suitable choice for process industry. This is the reason for applying PID-PLUS [1] in wired control of a wireless feedback plant. However, the approach is inadequate for wireless control application. This is because the fast update rate requirement imposed on the field wireless actuator can quickly shorten its battery lifetime. Therefore, the EWMA filter is proposed for use since it is computationally efficient and it is effective in addressing these issues. This is to preserve the traditional PID controller for wireless control of plants. Therefore, the application of an EWMA filter can be expanded from mainly semiconductor industries to process industries.

The key advantage of using an EWMA filter instead of an EWMA controller is that it does not require replacement of the conventional PID controller in plants, especially in oil and gas industries where most controllers are PID-based. Additionally, once the filter parameter is designed, the EWMA filter can be simply implemented in digital form on plant side. The implementation is much simpler compared to other techniques such as Kalman filter and Smith predictor.

In the next chapter, the existence of packet dropout in the wired link experiment will be described. In addition, the EWMA filter design steps will be presented as a guideline for addressing the delay and packet dropout issues in WNCS.

6

The Basics of an EWMA Filter

6.1 Introduction

It can be seen from Chapter 5, Section 5.2.4 that the applications of EWMA filters and controllers are not limited to only the semiconductor industry. Instead, they are also being used widely in other industries such as finance, the stock market, plant power load forecasting and so forth [125]. This chapter will cover some fundamental knowledge about the EWMA filter and its configurations. The main focus will be on the first- and second-order configurations. From a control perspective, the key advantage of using an EWMA filter instead of an EWMA controller is that it does not require replacement of the conventional PID controller system of process plants, especially in oil and gas or those industrial process plants in which most controllers are PID-based. Additionally, once a filter parameter is designed, the EWMA filter can be implemented in digital form on the process plant side. The implementation is much simpler compared to other approaches such as Kalman filter or Smith predictor.

6.2 First-order EWMA Filter

The typical use of EWMA filter is in the form of a first-order filter performing an operation called "single exponential smoothing" [59]. The analytical formula of the filter is shown below.

$$y_k = \alpha y_m + (1 - \alpha)y_{k-1} \qquad (6.1)$$

Here, y_k is the filter's output; y_m is the measurement value; y_{k-1} is the past filter's output; α is the filter's coefficient and its value is in the range: $0 < \alpha <= 1$. In the case $\alpha = 1$, there is no filtering to the signal. When α approaches 0, the measurement signal is discarded. The effect of the filter on a random signal given a specific value of α is presented in Figure 6.1.

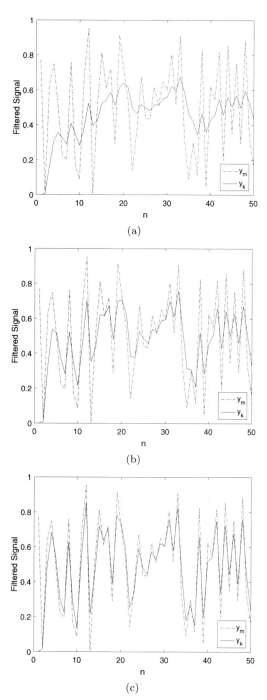

FIGURE 6.1: Filtering signal with 1st-order EWMA filter (a) $\alpha = 0.25$, (b) $\alpha = 0.50$, (c) $\alpha = 0.75$

In a particular case where $\alpha = 0.5$, i.e., Figure 6.1b, it can be said that the filter's weights are equally distributed to both the measurement value and the past filter's output. Therefore, the filter's output is the average of the measurement value and its past output: $y_k = (y_m + y_{k-1})/2$; and EWMA filter becomes 2-data-point moving average (MA) filter.

6.3 Second-order EWMA Filter

The second-order EWMA filter is also known as the double exponential filter. This is the combination of the two first-order filters into a cascaded one. A simplified version of the filter is described below [59].

$$y_k = \alpha^2 y_m + 2(1-\alpha)y_{k-1} - (1-\alpha)^2 y_{k-2} \qquad (6.2)$$

It is noted that this filter's application is limited due to the complexity to solve the equation with four unknown variables. Thus, often the simplification of the filter is used. The filter is then equivalent to two cascaded identical first-order EWMA filters. The overall filter's coefficient is α and its range varies from 0 to 1 like the first-order EWMA filter. The effect of the filter on a random signal given a specific value of α is presented in Figure 6.2. Compared to the first-order filter, the second-order filter has much slower response given same value of α, i.e., Figure 6.1a and Figure 6.2a when $\alpha = 0.25$.

An analysis of the performance of the 1st- and 2nd-order EWMA filters (based on MSE values) at different values of α is presented in Table 6.1.

TABLE 6.1: MSE comparison between first-order and second-order EWMA filters

α	1st-order EWMA	2nd-order EWMA
0.25	0.0660	0.0964
0.50	0.0393	0.0683
0.75	0.0202	0.0353

As seen from the table, when α is small MSE values are higher than when it is large. In general, at the same value of α, a 1st-order EWMA filter is more effective than a 2nd-order one because it results in lower MSE value, a common indicator representing the difference between the filtered signal and the original one.

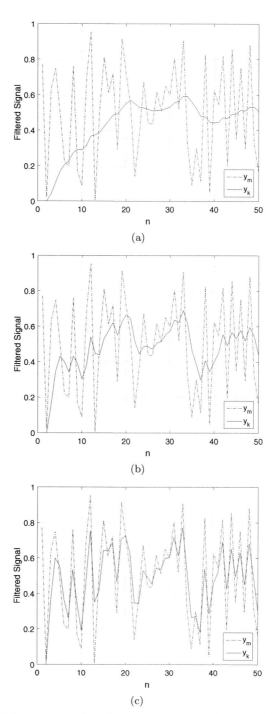

FIGURE 6.2: Filtering signal with 2nd-order EWMA filter (a) $\alpha = 0.25$, (b) $\alpha = 0.50$, (c) $\alpha = 0.75$

6.4 Higher-order EWMA Filter

Although the higher-order EWMA filter may well perform in certain signal smoothing applications, its use is limited since the high-order filter results in complexity when selecting its coefficients. Therefore, usually the first-order filter is used, and for some expanded applications, the second-order filter is in place.

6.5 Comparison with Other Filters

In both analog and digital fields, there are various filters available for selection such as the moving average filter (MA), Kalman filter, low-pass filter, high-pass filter, band-pass filter, and band-stop filter. Each of them has its own application areas such as for the semiconductor industry, banking and finance, and the military. The one that is closest to the EWMA filter is the MA filter. Thus, this section mainly focuses on comparison with the MA filter. Table 6.2 provides a summary of the comparison.

TABLE 6.2: Comparison of MA and EWMA filters

	MA Filter	**EWMA Filter**
No. of Data Point	Many (latest n points)	Most recent one point
Mechanism	Using an equal weight for every point	Different (exponential) weights for each point, placing more on recently measured value, less on past value
Effectiveness	Typically less effective	Typically more effective
Simplicity	Less simple, need to take average of recent values	More simple, use only one datum point

Figure 6.3 visualizes the performance comparison between first-order EWMA filter and MA filter (with number of averaging points $n = 5$).

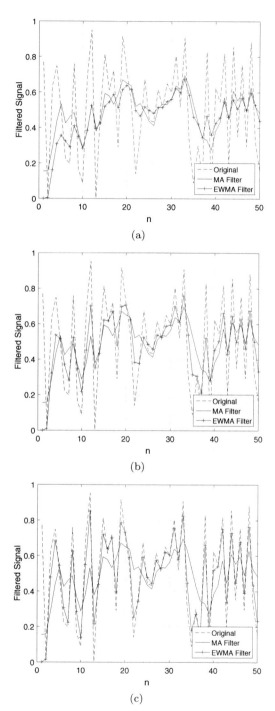

FIGURE 6.3: Comparison of MA and EWMA filters (a) $\alpha = 0.25$, (b) $\alpha = 0.50$, (c) $\alpha = 0.75$

As seen from the figure, when α increases, the EWMA filtered signal is closer to the original (unfiltered) signal; this helps to close the gap between them, while the MA filtered signal is smoother than that with EWMA when $\alpha = 0.25$. When $\alpha = 0.50$, both filters have similar signal levels and performance and when $\alpha = 0.75$, the EWMA filtered signal is better than the MA one. The difference in terms of MSE values is further detailed in Table 6.3.

TABLE 6.3: MSE comparison between first-order EWMA and MA filters

α	1st-order **EWMA**	**MA Filter (n=5)**
0.25	0.0660	0.0619
0.50	0.0393	0.0619
0.75	0.0202	0.0619

Through this example, it is seen that the similar or better level of performance can be achieved easily using the EWMA filter. The filter results in less computation overhead and it also consumes less memory compared to the MA filter. In this example, the EWMA filter uses only 2 data points: recent past and current ones while the MA filter uses 5 data points.

6.6   Summary

In this chapter, the fundamentals of the EWMA filter has been presented. Both first-order and second-order filters are discussed in detail. Although the higher-order filter has some particular applications, its usage on process control is limited. The following chapter will further discuss the advanced filter design for delay and packet dropout compensation.

7

Advanced Dual Purpose EMWA Filter Design for Delay and Packet Dropout Compensation

7.1 Introduction

This chapter describes the theoretical framework for designing the dual purpose EWMA (dpEWMA) filter for both delay and packet dropout compensation of wireless networked control systems (WNCSs). This is to improve the control performance of the wireless process plant by applying the filter at the plant's actuator input. A very typical wireless networked control system consisting of a controller, an actuator, a plant, a sensor node with complete control and feedback paths, shown in Figure 7.1 [172], is considered. In addition, the WirelessHART network is modeled with both upstream and downstream delay components. The overall numerical control performances (in terms of setpoint tracking and disturbance rejection) achieved via using the developed filter are discussed.

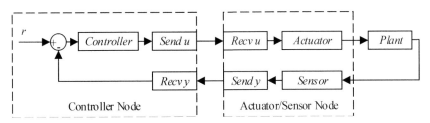

FIGURE 7.1: Typical WNCS model

As seen from the figure, WNCS is separated into three main blocks, namely controller node, actuator or sensor node and plant. The controller node receives setpoint (r) from the plant's operator, and feedback signal (y) from the plant's sensor, then issues a control signal (u) to the plant's actuator. The communication between the controller node and the actuator or sensor node is wireless. Although, it can be of protocols such as Wi-Fi or Bluetooth, this research mainly focuses on WirelessHART instead. This is because WirelessHART has special features supporting suitable communication in industrial environments. Therefore, the WNCS in this research can be named WirelessHART NCS or WHNCS.

Since there are always delays in communication between the controller and the actuator or sensor, for simplification purposes, the *Send u*, *Recv u*, *Send y*, and *Recv y* in the WNCS model are represented by delay components in Figure 7.2.

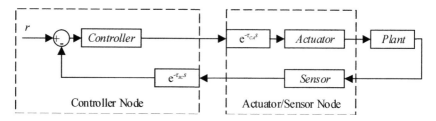

FIGURE 7.2: Simplified typical WNCS model

Here, it should be noted that the delay on the control path (τ_{CA}) is placed at the actuator node, while the delay on the feedback path (τ_{SC}) is placed at the controller node. This helps to represent well the nature of delays imposed on the receiver nodes in any typical communication system.

Further simplification of the WNCS model results in the representation of WHNCS in Figure 7.3. Firstly, the *Controller* in Figure 7.2 is represented by $C(s)$ in Figure 7.3. Secondly, a series of *Actuator*, *Plant* and *Sensor* blocks are represented by $P(s)$ which shows a process plant as a unified unit. Thirdly, the two delay components are used to represent the network induced delays in WHNCS. Finally, the reference signal is represented by $R(s)$ and the plant output is represented by $Y(s)$.

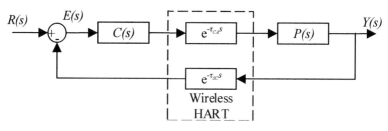

FIGURE 7.3: Symbolized typical WHNCS model

In order to design a dpEWMA filter, its position in the WHNCS should be first selected. Based on Figure 7.3, there are two main possible locations of the filter: (i) at controller node, on the control path or on the feedback path, and (ii) at actuator / sensor node, on the control path or on the feedback path. In this research, the location of the filter is selected to be at the actuator / the sensor node, on the control path (just prior to the *Actuator* block in Figure 7.2). This selection guarantees that the received signal can be filtered or processed before it is used at the plant. Therefore, in case the signal is disrupted, the filter can be applied and it will help help to improve the plant control performance. Thus, the WHNCS model with dpEWMA filter is presented in Figure 7.4 [2].

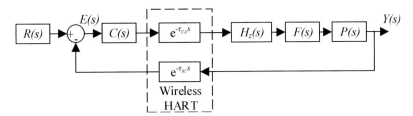

FIGURE 7.4: Typical WHNCS model with dpEWMA filter

Based on the figure, the WirelessHART network is modeled with the two delay components: upstream (from plant to controller) and downstream (from controller to plant). The plant is a sample system with a zero-hold order transfer function $H_z(s)$ placed at the plant input, while the dpEWMA filter is represented by $F(s)$.

In the later sections, the filter is first designed to handle controller delay. It is then adjusted to address packet dropout if it exists. The setpoint tracking and disturbance rejection performances of plants using the filter will be analyzed and consideration of the process with long delay is discussed.

7.2 dpEWMA Filter Design for Delay Compensation

7.2.1 Compensator Design

The delay compensator function used in this research is a function of an EWMA filter [173]. Its continuous form, identical to a first-order low-pass filter in (7.1), is used for filter design, while the discrete form in (7.2) is used to implement the filter given any plant.

$$F(s) = \frac{1}{\tau_F s + 1} \tag{7.1}$$

where $F(s)$ is the filter transfer function, τ_F is the filter time constant.

$$c(k) = \alpha \Re(k) + (1 - \alpha)c(k - 1) \tag{7.2}$$

where α is the compensator coefficient ($0 < \alpha < 1$), $\alpha = 1$ means no filtering and $\alpha \to 0$ means ignoring measurement value, $c(k)$ is the compensated control signal at the current time, $c(k - 1)$ is the most recent compensated control signal, $\Re(k)$ is the received control signal at the current time.

When $\triangle t$ is the sampling period, based on [173], the compensator coefficient α is defined as

$$\alpha = \frac{\triangle t}{\tau_F + \triangle t} \tag{7.3}$$

The control path transfer function of the system in Figure 7.4 is

$$G(s) = C(s)e^{-\tau_{CA}s}H_z(s)\frac{1}{\tau_F s + 1}P(s) \tag{7.4}$$

where $C(s)$ is the transfer function of the PID controller, $H_z(s)$ is the transfer function of the zero-order hold block, $P(s)$ is the transfer function of the process plant, τ_{CA} is the controller-to-actuator delay. In WNCS, it is also referred to as downstream delay (τ_d). In addition, the plant-to-controller delay τ_{SC}, referred to as upstream delay (τ_u), can be lumped with plant dead-time in $P(s)$.

The effect of downstream delay and delay compensator on the process plant is represented by (7.5).

$$E_F(s) = e^{-\tau_{CA}s}\frac{1}{\tau_F s + 1} \tag{7.5}$$

For a fast process, τ_{CA} is relatively small, close to 0, hence, the estimation in (7.6) holds.

$$e^{-\tau_{CA}s} \approx 1 - \tau_{CA}s \tag{7.6}$$

Equation (7.6) becomes

$$E_F(s) \approx \frac{-\tau_{CA}s + 1}{\tau_F s + 1} \tag{7.7}$$

The numerator has coefficient $\tau_{CA} > 0$, hence for a step response of (7.7), the response will begin from the negative region moving toward the positive region and approaching the setpoint, 1, as shown in Figure 7.5.

From the figure, when $\tau_F < \tau_{CA}$, the response is faster, thus stiffer and viceversa. In addition, the initial and final values of $E_F(s)$ can be found using initial and final theorems. In particular, $E_F(t = 0) = \lim_{s \to \infty}(sE_F(s)) = -\tau_{CA}/-\tau_F$ and $E_F(t = \infty) = \lim_{s \to 0}(sE_F(s)) = 1$. In the following steps, the filter design is presented to tackle the negative initial value of $E_F(t)$. Thus, the filter's performance does not depend on this initial value.

The analytical solution of (7.7) presented in [173] is rewritten as

$$E_F(t) = 1 - (1 + \frac{\tau_{CA}}{\tau_F})e^{-t/\tau_F} \tag{7.8}$$

To eliminate the effect of (7.7) on the system, in response to a step change,

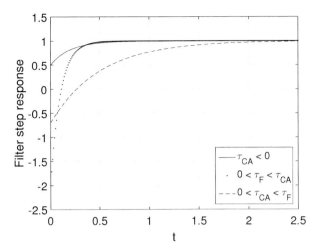

FIGURE 7.5: Step response of a delayed EWMA filter $E_F(s)$

$E_F(t)$ in (7.8) must reach setpoint equal to 1 within a certain period, e.g., t_m ($t \geq t_m$). Hence by setting the left-hand side (LHS) of (7.8) to 1, one has

$$(1 + \frac{\tau_{CA}}{\tau_F})e^{-t/\tau_F} \approx 0, \forall t \geq t_m \qquad (7.9)$$

By rewriting (7.9), it is found that

$$\frac{1 + \frac{\tau_{CA}}{\tau_F}}{e^{t/\tau_F}} \approx 0, \forall t \geq t_m \qquad (7.10)$$

For approximation purposes, with $a\%$ tolerance, (7.10) can be rewritten as

$$\frac{1 + \frac{\tau_{CA}}{\tau_F}}{e^{t/\tau_F}} \approx \frac{a}{100}, \forall t \geq t_m \qquad (7.11)$$

or

$$1 + \frac{\tau_{CA}}{\tau_F} \approx \frac{a}{100}e^{t/\tau_F}, \forall t \geq t_m \qquad (7.12)$$

At the lower bound condition, (7.12) can be rewritten as

$$1 + \frac{\tau_{CA}}{t_m}x \approx \frac{a}{100}e^x \qquad (7.13)$$

where

$$x = \frac{t_m}{\tau_F} \qquad (7.14)$$

Solving (7.13) gives a value of x, based on (7.14), τ_F can be found, thus the EWMA delay compensator is designed. The solution for (7.13) can be obtained using a computer program with a graphical approach. α is then calculated using (7.3) with the knowledge of process plant sampling time $\triangle t$. Therefore, the discrete form of EWMA delay compensator in (7.2) can be implemented. By eliminating the downstream delay in (7.7), response of the wireless control path is close to that without the downstream delay as described in (7.15). This is a standard form of a process control system.

$$G(s) \approx C(s)H_z(s)P(s) \qquad (7.15)$$

It is imperative to mention that this research does not focus on improving performance of any standard process control as described in (7.15). Instead, it improves the wireless plant's performance by mitigating the downstream delay effect.

7.2.2   Numerical Demonstration

Considering a fast plant with $\tau_{CA} = 0.2$ s, sampling at $\triangle t = 0.1$ s, it is desired that the plant is able to compensate for the delay before the next sample is carried out. This means, in response to a step change, $E_F(t)$ in (7.8) must reach setpoint equal to 1 within $t_m = 0.1$ s. From (7.13), with 1% tolerance, based on graphical solution of (7.13) shown in Figure 7.6, $x = 7.36$. Hence, based on (7.14), $\tau_F = 0.0136$ s. From (7.13), $\alpha = 0.8804$. The filter's response to step change at 0.1 s is shown in Figure 7.7.

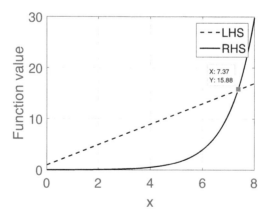

FIGURE 7.6: Graphical solution of (7.13)

As seen from the figure, the compensated filter $E_F(s)$ has a faster response compared to the original one. Its digital implementation follows similar response and is close to the setpoint. Here, the negative initial value does not exist for the digital filter. This is because the filter is designed to reach setpoint after $t_m = 0.1$ s. Even with the existence of the delay $\tau_{CA} = 0.2$ s, the digital filter's response is still faster than that of the original one. This means,

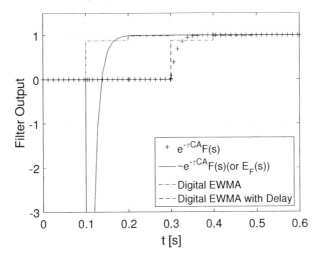

FIGURE 7.7: EWMA filter's response to step change at 0.1 s

the filter is able to compensate for the networked delay and its response is approximated to a unity (or $e^{-\tau_{CA}s}F(s) \approx 1$). This makes (7.15) valid.

7.3 dpEWMA Filter Design for Packet Dropout Compensation

7.3.1 Modified Compensator Design

Based on (7.2), it can be seen that the α parameter represents the weight placed on a successfully received control signal. Since packet dropout occurs, the control signal can be lost, and if the control signal is lost, the parameter's role in providing good current control signal is void. In this case, the other term with the weight $1 - \alpha$ plays an important role in providing current control signal. Hence, the parameter needs to be adjusted to accommodate this packet dropout rate and reduce its effect on the current measurement if packet dropout occurs. Not losing generality, it is assumed that at any instance, two consecutive scheduled control packets are sent to the plant's actuator. They have values of x and y, respectively. There are four possibilities with the state of these two packets: (0, 0), (0, 1), (1, 0), (1, 1) where 0 indicates the packet is received, while 1 indicates the packet is dropped (does not reach the actuator). For demonstration purposes, packet receiving rate value is associated with these particular two packets only.

With non-zero initial condition $\delta > 0$, based on (7.2), the values that the actuator receives are

- Case $(0, 0)$: both packets are received, $p_r = 100\%$.

$$c(1) = \alpha x + (1 - \alpha)\delta$$
$$c(2) = \alpha y + (1 - \alpha)(\alpha x + (1 - \alpha)\delta)$$

(7.16)

- Case $(0, 1)$: packet x is received, packet y is dropped, $p_r = 50\%$.

$$c(1) = \alpha x + (1 - \alpha)\delta$$
$$c(2) = (1 - \alpha)(\alpha x + (1 - \alpha)\delta)$$

(7.17)

- Case $(1, 0)$: packet x is dropped, packet y is received, $p_r = 50\%$.

$$c(1) = (1 - \alpha)\delta$$
$$c(2) = \alpha y + (1 - \alpha)(1 - \alpha)\delta$$

(7.18)

- Case $(1, 1)$: both packets are dropped, $p_r = 0\%$.

$$c(1) = (1 - \alpha)\delta$$
$$c(2) = (1 - \alpha)(1 - \alpha)\delta$$

(7.19)

In the first three cases, though there is partial packet loss, part of the scheduled control signals still can be executed. Differently, in the fourth case, the executed control signals heavily depend on the initial control signal condition. In control engineering, the control signal is issued to maintain stability of the process plant. However, based on (7.19), if consecutive signals are lost, control signal values will decrease over time, causing instability of states of the process plant. Therefore, when the system has packet dropout, this research proposes an adjustment of the filter parameter to reflect the packet receiving rate (probability). The new value of this parameter is then being calculated as

$$\theta = \alpha p_r$$

(7.20)

where θ is adjusted receiving weight; α is receiving weight found in Section 4.3; p_r is packet receiving rate.

This parameter depends on two factors: initial receiving weight, and packet receiving rate. The adjustment is only applied when there is existence of packet dropout. Equation (7.2) becomes

$$c(k) = \theta \Re(k) + (1 - \theta)c(k - 1)$$

(7.21)

Thus, the values the actuator receives are:

- Case $(0, 0)$: both packets are received, $p_r = 100\%$.

$$c(1) = p_r \alpha x + (1 - p_r \alpha)\delta$$
$$c(2) = p_r \alpha y + (1 - p_r \alpha)(p_r \alpha x + (1 - p_r \alpha)\delta)$$

(7.22)

- Case (0, 1): packet x is received, packet y is dropped, $p_r = 50\%$.

$$c(1) = p_r\alpha x + (1 - p_r\alpha)\delta$$
$$c(2) = (1 - p_r\alpha)(p_r\alpha x + (1 - p_r\alpha)\delta)$$
(7.23)

- Case (1, 0): packet x is dropped, packet y is received, $p_r = 50\%$.

$$c(1) = (1 - p_r\alpha)\delta$$
$$c(2) = p_r\alpha y + (1 - p_r\alpha)(1 - p_r\alpha)\delta$$
(7.24)

- Case (1, 1): both packets are dropped, $p_r = 0\%$.

$$c(1) = (1 - p_r\alpha)\delta$$
$$c(2) = (1 - p_r\alpha)(1 - p_r\alpha)\delta$$
(7.25)

Different from (7.19), (7.25) takes into consideration packet receiving rate. Thus, in the worst case, if $p_r = 0\%$, control signals applied to the actuator are maintained as the initial condition δ. This helps to maintain the recent stable state of the process plant in the event of packet dropout.

7.3.2 Numerical Demonstration

Following up the numerical demonstration in Section 7.2.2, Figure 7.8 presents the responses of filter and modified filter when packet dropout is 20%. As seen from the figure, both filtered and modified filtered signals have lower magnitudes as compared to the original received signals. This is because the filter coefficients are smaller than unity, thus, only part of the received signals is used. In addition, the modified filter has a response slower than that of the unmodified one to reduce the effect of packet dropout on the scheduled control signals generated by the filter. Therefore, it is suitable for maintaining the recently used control signals longer.

7.4 dpEWMA Filter's Performance

7.4.1 Setpoint Tracking

The key advantage of an EWMA filter is the ability to maintain a stable control signal. This characteristic itself carries a disadvantage during the setpoint change. This section elaborates on the dragging effect of the filter on control system performance.

Consider a series of control signals sent to a process plant as depicted in Figure 7.9. In this figure, for each received control signal, the EWMA filter will

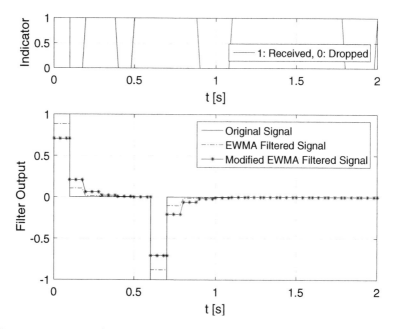

FIGURE 7.8: EWMA and dpEWMA response in presence of 20% packet dropout

generate a scheduled control signal. Thus, using (7.2), one has the following relationship.

$$c(k-1) = \alpha\Re(k-1) + (1-\alpha)c(k-2) \tag{7.26}$$

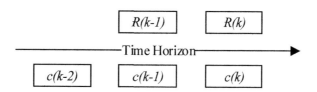

FIGURE 7.9: Received ($\Re$) and scheduled (c) control signals without disturbance

Not losing generality, it is possible to assume that during the setpoint change, the sent control signal has greater magnitude than the previous scheduled control signal. This means $\Re(k-1) > c(k-2) > 0$. The inequality can be expressed as

$$\Re(k-1) = c(k-2) + \triangle(k-2) \tag{7.27}$$

or

$$\frac{\Re(k-1)}{c(k-2)} = 1 + \triangle(k-2) \tag{7.28}$$

where $\triangle(k-2) > 0$. Hence, (7.26) becomes

$$c(k-1) = \alpha\triangle(k-2) + c(k-2) \tag{7.29}$$

or

$$\frac{c(k-1)}{c(k-2)} = 1 + \alpha\triangle(k-2) \tag{7.30}$$

Since $\alpha < 1$ as described in Section 7.2, by comparing (7.28) and (7.30), one obtains $0 < c(k-1) < \Re(k-1)$. This means the scheduled control signal has smaller magnitude as compared to the received control signal. During the rise time introduced by the positive setpoint change, the plant response will be slower than when not using the filter. Additionally, the increment or decrement rate of the scheduled control signal will be smaller than that of the received control signal. Thus, the system with the filter will exhibit longer rise time, and slower response to on-demand change. The overshoot, if it exists, will be increased because at peak response, the reaction to the change in control signal is slightly dragged due to the effect of the filter coefficient and the past scheduled control signal. To address this issue, the past scheduled control signal in (7.2) should be eliminated during the setpoint change, hence, the modified scheduled control signal is

$$c(k) = \alpha\Re(k) \tag{7.31}$$

Here, (7.31) can be referred to as a partial EWMA filter since the past scheduled control signal is not accounted for.

Figure 7.10 presents a demonstration of setpoint performance of different EWMA filter configurations. The plant under control is a first-order plant with transfer function of $\frac{1}{s+1}$ while the controller is PI with parameters of $K_P = 11.360$, $K_I = 37.971$. The coefficients of the filter and the modified filter are 0.8804 and 0.7924, respectively (see Section 7.2.2). As seen from the figure, the plant with fully configured filters is faster in both response and settling as compared to the plant with PI only. In addition, the plant with modified filter introduces the highest percent overshoot which is due to the accumulated effect of the past control signals. Differently, when the filters are half configured they can ignore the past control signals during the setpoint change. Thus, the plant response is comparable (insignificant overshoot and settling time differences, slightly slower in response) to that with PI only.

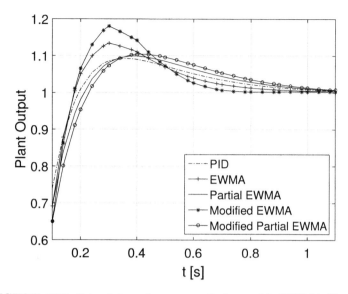

FIGURE 7.10: Setpoint performance of plant with EWMA filters

7.4.2 Disturbance Rejection

It has been detailed in Section 7.3 that the EWMA filter can preserve well the stable control signal. Therefore, it can be effective for disturbance rejection as described in what follows. When disturbance occurs at the input of a plant, instead of receiving only the control signal sent from the controller, the EWMA filter will receive additional disturbance as shown in Figure 7.11.

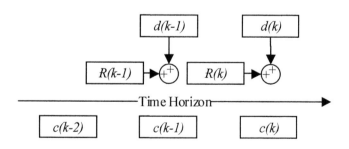

FIGURE 7.11: Received ($\Re$) and scheduled control (c) signals with disturbance (d)

Based on the figure and (7.2), the scheduled control signal subject to disturbance at instances $(k-2), (k-1)$ and (k) are given in (7.32), (7.33), and

(7.34), respectively.

$$c(k-2) = \alpha(\Re(k-2) + d(k-2)) + (1-\alpha)c(k-3) \qquad (7.32)$$
$$c(k-1) = \alpha(\Re(k-1) + d(k-1)) + (1-\alpha)c(k-2) \qquad (7.33)$$
$$c(k) = \alpha(\Re(k) + d(k)) + (1-\alpha)c(k-1) \qquad (7.34)$$

For generalization, (7.34) is rearranged as

$$\begin{aligned} c(k) &= \alpha\Re(k) \\ &+ \alpha d(k) \\ &+ (1-\alpha)^1 c(k-1) \end{aligned} \qquad (7.35)$$

Substituting (7.33) into (7.35), one has

$$\begin{aligned} c(k) &= \alpha\Re(k) \\ &+ \alpha d(k) \\ &+ (1-\alpha)^1(\Re(k-1) + d(k-1)) + (1-\alpha)^2 c(k-2) \end{aligned} \qquad (7.36)$$

or

$$\begin{aligned} c(k) &= \alpha\Re(k) + \alpha(1-\alpha)^1\Re(k-1) \\ &+ \alpha d(k) + \alpha(1-\alpha)^1 d(k-1) \\ &+ (1-\alpha)^2 c(k-2) \end{aligned} \qquad (7.37)$$

Substituting (7.32) into (7.37) one has

$$\begin{aligned} c(k) &= \alpha\Re(k) + \alpha(1-\alpha)^1\Re(k-1) \\ &+ \alpha d(k) + \alpha(1-\alpha)^1 d(k-1) + \alpha(1-\alpha)^2(\Re(k-2) + d(k-2)) \\ &+ (1-\alpha)^3 c(k-3) \end{aligned} \qquad (7.38)$$

or

$$\begin{aligned} c(k) &= \alpha\Re(k) + \alpha(1-\alpha)^1\Re(k-1) + \alpha(1-\alpha)^2\Re(k-2) \\ &+ \alpha d(k) + \alpha(1-\alpha)^1 d(k-1) + \alpha(1-\alpha)^2 d(k-2) \\ &+ (1-\alpha)^3 c(k-3) \end{aligned} \qquad (7.39)$$

Based on (7.35), (7.37), (7.39), generalizing the formula for current scheduled control signal (at instance k) with reference to the past n scheduled control signals gives

$$\begin{aligned} c(k) &= \alpha\Re(k) + \alpha(1-\alpha)^1\Re(k-1) + ... + \alpha(1-\alpha)^{n-1}\Re(k-n+1) \\ &+ \alpha d(k) + \alpha(1-\alpha)^1 d(k-1) + ... + \alpha(1-\alpha)^{n-1} d(k-n+1) \\ &+ (1-\alpha)^n c(k-n) \end{aligned} \qquad (7.40)$$

The effect of disturbance on the current scheduled control signal is expressed as

$$\begin{aligned} D(k) &= \alpha d(k) \\ &+ \alpha(1-\alpha)^1 d(k-1) \\ &+ ... \\ &+ \alpha(1-\alpha)^{n-1} d(k-n+1) \end{aligned} \qquad (7.41)$$

Although the current scheduled control signal is affected by not only the current disturbance $d(k)$ but also the past disturbances $d(k-1), ..., d(k-n+1)$ as in (7.41), the effect of the past disturbances quickly depreciates as the number of considered disturbances increases. This is because $0 < \alpha < 1$, thus, $0 < 1 - \alpha < 1$ and $(1-\alpha)^{n-1} \to 0$ after only a few values of n especially when $\alpha > 0.5$ (see Table 7.1). It should be noted that this selection of $\alpha > 0.5$ is made to ensure the received control signal receives significant weightage else the system response will be stagnant.

TABLE 7.1: Example of disturbance weights in EWMA filters

α	$1 - \alpha$	n	$(1-\alpha)^{n-1}$	Disturbance Weight $\alpha(1-\alpha)^{n-1}$
0.6	0.4	1	1	0.6
0.6	0.4	2	0.16	0.096
0.6	0.4	3	0.064	0.0384
0.6	0.4	4	0.0256	0.01536
0.8	0.2	1	1	0.8
0.8	0.2	2	0.04	0.032
0.8	0.2	3	0.008	0.0064

In Table 7.1, it is seen that when $\alpha = 0.6$, there are only two significantly effective past disturbances with respect to $n = 2, 3$ as the third one has close to 0 weight. While when $\alpha = 0.8$, the first effective disturbance already has the weight close to 0 which is 0.032 when $n = 2$.

Therefore, when mean disturbance $\bar{d}(k, n) = \frac{1}{n} \sum(d(k - i))$ is considered, the mean effect of disturbance on the scheduled control signal in (7.41) is approximated as

$$D(k) = (\alpha + \alpha(1 - \alpha)^1 + ... + \alpha(1 - \alpha)^{n-1})\bar{d}(k, n) \qquad (7.42)$$

Hence, the accumulated weight for the disturbance is

$$w_D(k) = \alpha + \alpha(1 - \alpha)^1 + ... + \alpha(1 - \alpha)^{n-1} \qquad (7.43)$$

Here, it is interesting to analyze the sensitivity of the disturbance weight on the filter parameter α. The sensitivity function of $w_D(k)$ is described as

$$
\begin{aligned}
& S_{w_D(k)/\alpha} \\
&= \frac{\alpha}{w_D(k)} \frac{\delta w_D(k)}{\delta \alpha} \\
&= \frac{\alpha[1 + (1 - \alpha) + \alpha(-1) + ... + (1 - \alpha)^{n-1} + \alpha(n - 1)(1 - \alpha)^{n-2}(-1)]}{\alpha + \alpha(1 - \alpha)^1 + ... + \alpha(1 - \alpha)^{n-1}} \\
&= \frac{1 + (1 - \alpha) + \alpha(-1) + ... + (1 - \alpha)^{n-1} + \alpha(n - 1)(1 - \alpha)^{n-2}(-1)}{1 + (1 - \alpha)^1 + ... + (1 - \alpha)^{n-1}}
\end{aligned}
$$

$$(7.44)$$

As α approaches unity, $1 - \alpha$ approaches 0, and as α approaches 0, $1 - \alpha$ approaches 1, thus

$$
\begin{aligned}
S_{w_D(k)/\alpha} &\underset{\alpha \to 1}{=} 1 \\
S_{w_D(k)/\alpha} &\underset{\alpha \to 0}{=} 1
\end{aligned}
\tag{7.45}
$$

Based on (7.45), near α limits, the change in α results in the change in $w_D(k)$ with the ratio of 1. This means the disturbance weight is proportional to the filter coefficient with the ratio of 1. Since the filter coefficient is selected to be less than unity, the disturbance's effect is reduced by the amount of close to $1 - \alpha$ even if the previous disturbances are taken into consideration. Based on Table 7.1, Table 7.2 illustrates this disturbance reduction effect.

TABLE 7.2: Example of disturbance reduction in EWMA filters

n	Dist. Weight $\alpha(1-\alpha)^{n-1}$	Accum. Dist. Weight $\sum \alpha(1-\alpha)^{n-1}$	Act. Dist. Reduc. $1 - \sum \alpha(1-\alpha)^{n-1}$	Approx. Dist. Reduc. $1 - \alpha$
		$\alpha = 0.6$		
1	0.6	0.6	0.4	0.4
2	0.096	0.696	0.304	0.4
3	0.0384	0.7344	0.2656	0.4
4	0.01536	0.74976	0.25024	0.4
		$\alpha = 0.8$		
1	0.8	0.8	0.2	0.2
2	0.032	0.832	0.168	0.2
3	0.0064	0.8384	0.1616	0.2

For demonstration on the effect of the filter subject to plant input disturbance, the filters' responses to disturbances are presented in Figure 7.12. Based on Section 7.2.2, the coefficients of the EWMA filter and its modified version are 0.8804 and 0.7924, respectively. The sample time is set to 0.02 s for demonstration purposes.

It is seen from the figure that the filter helps to reduce disturbance's effect on control performance of the system. In particular, the filtered control signals follow well the disturbances and they settle to stable states after only three sampling instants for all cases. Therefore, disturbance's effect on control signal can be effectively and quickly eliminated by using the designed EWMA filter as described in Table 7.2 (see the case for $\alpha = 0.8$).

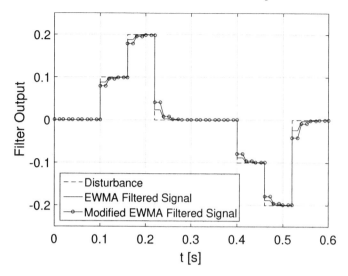

FIGURE 7.12: EWMA and dpEWMA responses in presence of disturbance

7.5   dpEWMA Filter for Process with Long Delay

In the previous sections, the filter design makes use of the controller delay (less than 1 s) which is only part of the overall network round-trip delay. However, the delay measured from the system may be longer than this value. Therefore, this section presents an approach for handling the system with long delays (exceeding 1 s).

It is seen that the approximation in (7.4) is valid for a small delay component close to 0. In control system engineering, it is always possible to express the unit of delay used in the system in terms of seconds (most commonly used), minutes, and hours. If the delay is longer than 1 s, it can be converted to minutes to make the value close to zero. A similar approach can be applied for delay longer than 1 minute or hour. Therefore, the filter time constant in (7.5) can be expressed with the same unit as the long delay. Finally, the sampling period in (7.4) should be in the same unit as the long delay as well.

In the later sections, the filter is first designed to handle controller delay. It is then adjusted to address packet dropout if it does exist.

7.6   Summary

It is seen from this chapter that the advanced EWMA filter design approach presented here helps to significantly improve process control performance. The design takes into consideration both delay and packet dropout which are key elements that affect control performance of a system. In the following chapters, the filter's performance over a wired link and wireless links is demonstrated.

8

The Filter's Performance over Wired Links

8.1 Introduction

This chapter discusses and analyzes results of this book. It starts with the measured network induced delays using end-to-end and one-end approaches. Second, the experiment results on wired link contention are presented. Third, control performance assessment for handling wired contention using the EWMA filter is discussed. Finally the assessment results using the developed real-time simulator WH-HILS are analyzed.

8.2 Network Induced Delay Measurement

Network induced delays collected through experiments are presented in this section. Both end-to-end and one-end delay measurement results are covered as each has its own importance. In particular, end-to-end delay measurement can be used to introduce a delay compensation mechanism before a message is sent to the destination while one-end delay can be used for overall network and controller optimization.

8.2.1 End-to-end Delay Measurement

8.2.1.1 Upstream Delay

For Publish measurement delays in Figure 8.1, it can be seen that, although the delay ranges between 2 and 3 s, some delays go beyond this range, up to 5 to 6 s. This is due to the overflow message queue in the node during the experiment. Therefore, the sample period of the WirelessHART node should not be shorter than the maximum measured delay in order to prevent the occurrence of overflow in the device's message queue.

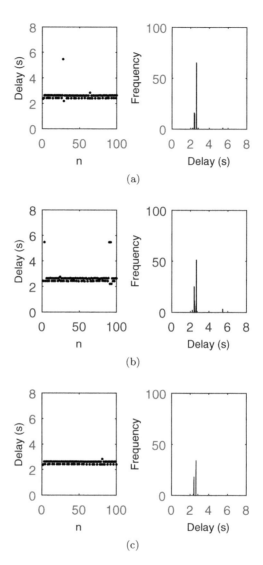

FIGURE 8.1: Upstream delay from node to gateway in Publish mode (a) low priority, (b) medium priority, (c) high priority

The results of Event communication mode are presented in Figure 8.2. As seen from the figure, the frequency of delay in the range of 2 to 3 s remains the highest. For Low priority, the delay is bounded below 3 s while at some instances, the delay reaches up to between 5 and 6 s. This is due to the nature of Event mode which informs the control system about the event happening in the plant. Basically, Event mode receives higher focus as compared to Publish mode. Thus, when multiple events are sent continuously to the gateway, certain events receive higher priority as compared to the rest. This can result in local contention at the gateway, especially with Medium and High priority status.

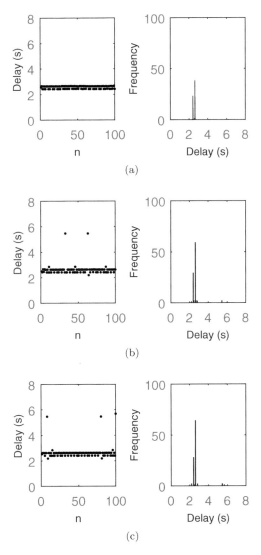

(a)

(b)

(c)

FIGURE 8.2: Upstream delay from node to gateway in Event mode (a) low priority, (b) medium priority, (c) high priority

Figure 8.3 represents the measurements of Maintenance mode with priorities from Low to High. This mode is dedicated to sending configuration messages in the network. Often, the payload in addition to its original length contains several other setting fields. Therefore, the longer the length of the payload, the longer the induced delay. The delay in this mode experiences more variation, though the delay is still in the range of 2 to 3 s. The frequency of delays longer than 5 s is higher for High priority. Sometimes, it takes up to 8 s to process a message completely.

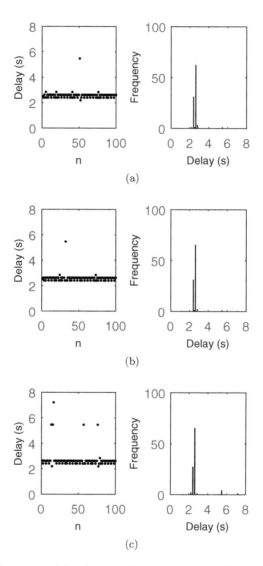

FIGURE 8.3: Upstream delay from node to gateway in Maintenance mode (a) low priority, (b) medium priority, (c) high priority

The Block Transfer mode is dedicated for transferring data from one node to another node. In this case, the node will serve the role of a relay which transmits data from source node to destination node. Therefore, the node does not need to consider the content of the message payload. Communication between this node and the destination must be established once it receives a command from the source. Hence, expected delay for this mode will be longer than the other cases.

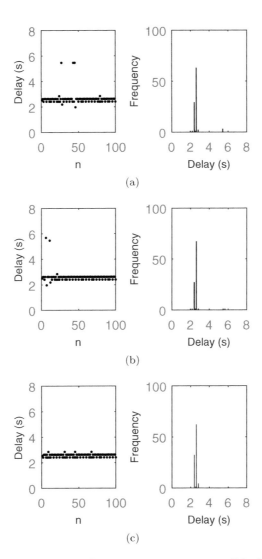

FIGURE 8.4: Upstream delay from node to gateway in Block Transfer mode (a) low priority, (b) medium priority, (c) high priority

However, in this case, the node directly sends the message to the gateway. Therefore, the delay remains within the range of 2 to 3 s (see Figure 8.4). It should be noted that, at Medium priority, the node experiences high delay initially, while at Low priority, it has some delays up to 5 to 6 s.

8.2.1.2   Downstream Delay

The average delay recorded for downstream communication is slightly lower than that captured for the upstream communication, i.e., 2.3 s. Here, the delay patterns for the gateway shown in Figure 8.5 are more stable and smoother as compared to the delay patterns recorded by the node. This is due to the less computational capacity of the node as compared to that of the gateway. Additionally, the wired data transfer rate between the node and the micro-computer (i.e., 9,600 kbps) is also less than that of the gateway with the computer (i.e., 115,200 kbps). At higher communication speed, the gateway is faster in sending data to the computer, thus it is more stable and less time-consuming.

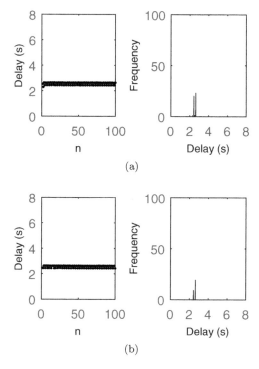

FIGURE 8.5: Downstream delay from node to gateway in Block Transfer mode (a) low priority, (b) high priority

In summary, in this experiment, the delays for both upstream and downstream communications are within the range of 2 to 3 s for all cases. This implies that the update period limit for the node or the gateway should not be less than 3 s. Update periods less than 3s will result in unpredictable delay in the network due to the overflow of message queue at either node or gateway.

8.2.1.3 Delay Summary

The summarized average upstream and downstream delay measurements are presented in Figure 8.6.

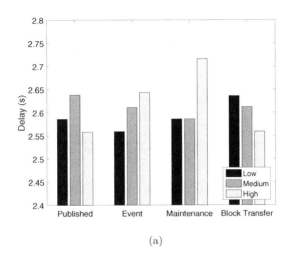

(a)

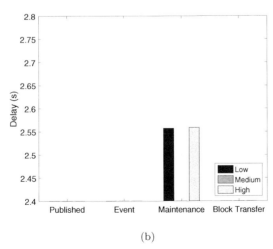

(b)

FIGURE 8.6: Average measured (a) upstream and (b) downstream delays

As discussed in Section 4.2.2, the node supports four upstream communication modes to the gateway. They are Publish, Event, Maintenance, and Block Transfer. Each has three priority levels: Low, Medium and High. For Publish mode, the Medium priority gives the highest delay. For Event mode, with increase in priority level, the delay also increases. For Maintenance mode, the same delay is recorded for both Low and Medium priorities while for High priority, the delay is higher at about 0.1 s. For Block Transfer mode, the delay decreases as the priority increases. On the other hand, only Maintenance mode is supported by the gateway for downstream communication to the node. The Maintenance mode has two priority levels: Low and High. As seen from Table 8.1, there is insignificant delay difference between them.

TABLE 8.1: Average measured delays

Priority	Low	Medium	High
Node to Gateway			
Publish (s)	2.586±0.309	2.638±0.512	2.558±0.106
Event (s)	2.559±0.106	2.611±0.424	2.643±0.522
Maintenance (s)	2.586±0.313	2.586±0.308	2.716±0.736
Block Transfer (s)	2.636±0.512	2.612±0.440	2.559±0.116
Gateway to Node			
Maintenance (s)	2.557±0.098		2.559±0.094

8.2.2   One-end Delay Measurement

The upstream and downstream delays measured over a 3-hour experiment with a single hop WNCS is presented in Figure 8.7. It is observed that the measured downstream delay is a constant value at 1.28 s while the upstream delay is time varying. As explained by the manufacturer, Linear Technology, the downstream delay measurement is based on only network scheduling for communication with the node from the gateway. Thus, if there is no change in network schedule, there is no change in the downstream delay. While for upstream delay the applied concept is different. The upstream delay is depending on the actual communication from the node to the gateway, and this is time-varying as each time the node sends a message to the gateway, the communication delay is different from another. Therefore, the upstream delay measurement from the gateway utilizes all measured delays from the node to the gateway during a specific period defined by the manufacturer. The upstream delay measured in this experiment has the following statistics information: average delay is 1.626 s, standard deviation is 0.363 s, maximum delay is 4.263 s, and minimum delay is 1.160 s. This information is summarized in Table 8.2.

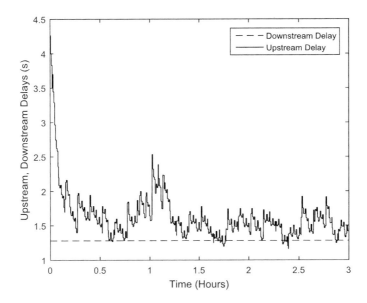

FIGURE 8.7: One-end delay measurement

TABLE 8.2: One-end measured delay statistics

Parameters	Downstream Delay	Upstream Delay
Average	1.280	1.626
Standard Deviation	0	0.363
Max	1.280	4.263
Min	1.280	1.160

8.2.3 Variable Payload and Delays

The sample generated control message payloads used in the experiment are presented in Table 8.3. It is shown that the payloads are randomized and represent well the nature of time-varying control payload when sending control actions from the controller to the plant.

TABLE 8.3: Sample generated control message payload

Payload Length (Bytes)	Sample Generated Payload (Hex)
1	ab
10	855e3b95f5c555dec0b6
20	d35badf5745132b7f87fb7e169c31873c4523279
30	88ba311c01483097fb1bf7224912d339bad419af e1b2a747a708b71738ad
40	902dec5eee650d10f7e856cdf00f5a5c182a22f1 2c4a23f343b7b1710cbfa3613eb6c4eaf75a64f8
50	563864350c30370af6fc3cdbb1263ad4e95652ea c6d400b898a8b23f9ac8f3d6b0628ec9d71d10c0 feaa9d66038d0f42a4be
60	b642850736aa34e006662125cc86506dff966db5 5a217877e3216c858d261688677b70671faf47f3 b7c2d01695680f4719412c06c9a463af0d383653
70	72f4ed2b075954e8f11564993a81d944227de0c3 53bd458d373ecff5a34faf365f50275206ebb9fa fde927a091e715cdd1b78149d707ced97847de00 4dd2969824f46fcf838d
78	49d8a57073b9d7e6274b49f79345c641deb1f5b0 54e72983cc7ca7a3ae80fde648e8b3f50109504c 71e0e91d60e2721b15c0db5af2e44c9eac775da4 219431cefe8368becb2069d830deebb4eb09

Figure 8.8 presents the resultant measured control delay with several payload lengths. The average values of delays are in the range from 2.55 s to 2.65 s. The payloads with lengths of 1 byte and 50 bytes exhibit the lowest standard deviation while the payload with length of 10 bytes exhibits the highest standard deviation. There is a decreasing trend in the standard deviation of the payload with length of from 10 bytes to 78 bytes, except for the case of 50 bytes. Overall, the control delay from the gateway to the node is 2.0 s to 3.3 s. Although, the average delay pattern for the payload of 10 to 30 bytes is similar to the pattern for that of 60, 70, 78 bytes (reducing both average delay and standard deviation), the dependence of control delay on packet length is insignificant.

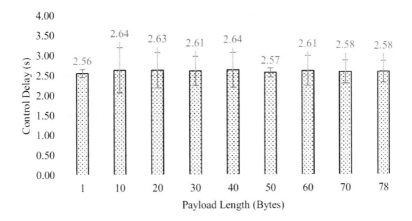

FIGURE 8.8: Measured control delay with varying payload length

A further look into the measured delay reveals that at some particular instance, the measured delay exceeds the nominal range of control delay. This results in high standard deviation for the message lengths of 10 bytes, 20 bytes, 30 bytes, 40 bytes, 60 bytes, 70 bytes, and 78 bytes in Figure 8.8.

After removing the outliers (defined in this book as those with extremely high delay or higher than 5 s) the measured control delay is replotted in Figure 8.9.

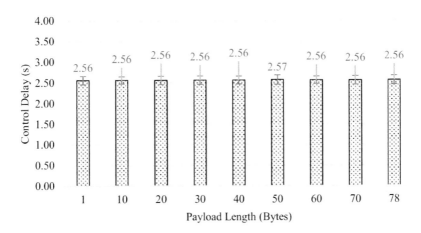

FIGURE 8.9: Measured control delay with varying payload length without outliers

From Figure 8.9, the control delay is close to a constant value of 2.56 s with similar standard deviation of close to 0.1 s. The detail of standard deviation for each case in Figure 8.8 and Figure 8.9 is presented in Table 8.4.

TABLE 8.4: Average measured delays' standard deviations

Payload Length (Bytes)	Std. Dev. in Figure 8.8 (s)	Std. Dev. in Figure 8.9 (s)
1	0.10	0.10
10	0.57	0.09
20	0.44	0.10
30	0.36	0.10
40	0.44	0.09
50	0.11	0.11
60	0.37	0.09
70	0.29	0.10
78	0.27	0.11

8.3   Wired Link Contention Experimental Results

Being an essential part of a WNCS, study of wired contention is important for assessing performance of the system. This is because wired communication's reliability can affect overall wireless control performance especially when packet dropout occurs. This section will describe the experimental results on wired link contention between a controller and a wireless node.

8.3.1   Wired Link Delay Measurement

Over a wired link communication, the main delay component is the signal processing time. This is because, even at the slowest communication speed of 9600 bps, ideally, the transmission time for a typical two-byte data (more than sufficient to represent value of a process parameter) is only 1.67 ms. In this case, the processing time is the time taken for the controller to prepare a control message to send to the WirelessHART node. It is equal to the time taken for the controller to initialize HDLC frame, construct a message payload to be sent, calculate FCS, assemble HDLC, FCS and pad both ends with flag sequences [2, 117]. The whole process is presumed to take up the amount of time equal to Pre-FCS, FCS and Post-FCS calculation times presented in Figure 8.10. The most time-consuming task is to calculate FCS, taking 0.1335 s. Pre-FCS calculation takes 0.0116 s and involves several tasks such as initialize HDLC frame, FCS polynomial pattern, remove flag sequence, byte-unstuff the frame, extract the received FCS and frame payload. After FCS is calculated, it is checked with the received FCS. If both are matched, the frame payload is considered a valid received frame. Otherwise it is an invalid frame. The verification takes 0.0024 s to be completed. Thus, the overall controller delay is much greater than the longest ideal wired transmission delay $\tau_{CA} = 0.1475s = 147.5ms >> 1.67ms$.

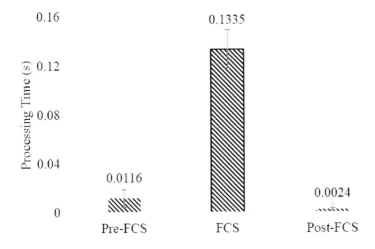

FIGURE 8.10: Processing time for validating HDLC frame

8.3.2 Wired Link Delay and Packet Dropout Correlation

Figure 8.11 is presented to illustrate the correlation between packet receiving rate (accurate response) and added delay. Here, added delay is the time difference between any two consecutive messages. When it is close to 0, the messages are sent continuously without a resting period. From the figure, when added delay is from 0.10 s, accurate response rate increases significantly from as low as 0.1% to as high as above 85% and remains consistent above this value. When there is an added delay, there is time for the node to respond to the request sent from the controller. This means, when the added delay is 0 s, or when the controller concurrently requests information from the node, the node suffers from correctly responding to the requests. Contention happens due to collision of concurrent packets sent to the node which is a cause for wired packet dropout. The extremely low accurate response rate (i.e., 0.1%) best describes the occurrence of contention. When the added delay is set to 0.40 s, the highest accurate response rate is achieved (99.5%).

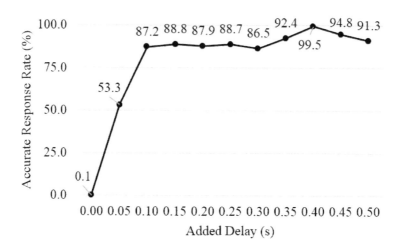

FIGURE 8.11: Average accurate response rate versus added delay

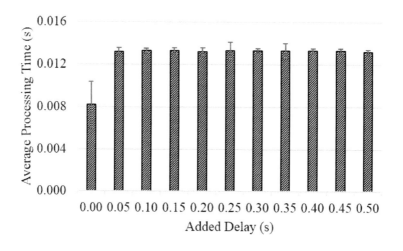

FIGURE 8.12: Average execution time versus added delay

Moreover, since the wired link delay is due to execution time by the controller (transmission time between controller and the node), Figure 8.12 shows the controller's execution time versus the added delay.

In addition, Table 8.5 shows its corresponding standard deviation for a complete two-way wired communication between the node and the controller. The execution time depends on the number of online tasks handled by the controller during the experiment period. The larger the number of tasks being run, the longer the time taken to complete one WirelessHART execution. On average, the execution time stays close to 0.013 s per execution. The standard deviation remains from 0.0002 s to 0.0004 s while for a particular case when there is no added delay, the standard deviation is excessively large, standing at 0.0022 s. This is due to continuous requests sent to the node that cause local contention and result in packet collision and dropout.

TABLE 8.5: Execution time's standard deviation

Added Delay (s)	Processing Time's Std. Dev. (ms)
0.00	2.2
0.05	0.4
0.10	0.2
0.15	0.3
0.20	0.4
0.25	0.8
0.30	0.2
0.35	0.7
0.40	0.2
0.45	0.2
0.50	0.2

Basically, the communication delay (execution time by controller) should be added to the controller's delay; however, due to the significantly small value of the delay (i.e., 0.013 s from Figure 8.12) as compared with the delay caused by preparing HDLC frame for the communication (i.e., 0.1475 s from Figure 8.10), this delay is considerably negligible and will not be taken into account for total controller delay. Since the estimated local controller's delay is $\tau_{CA} = 0.1475$ s, based on Figure 8.12, one can find: $p_{r1} = 87.20\%$, $p_{r2} = 88.80\%$, $t_{d1} = 0.10s$, $t_{d2} = 0.15s$, $t_d = \tau_{CA} = 0.1475s$. By using interpolation process as described in (7.11), the estimated accurate response rate is $p_r = 87.71\%$. Hence, based on (7.12), the packet dropout rate is $p_d = 12.29\%$.

To this point, it can be seen that it is important to have a packet dropout estimation algorithm that can be used by the plant to provide knowledge of previous measurements. The algorithm shall help to determine packet dropout rate more precisely compared to the aforementioned typical two-point estimation approach. Section 8.3.3 will detail a procedure for packet dropout estimation using an ARX-based Kalman filter approach. Furthermore, Section 8.4 will study the effect of controller's wired packet dropout and delay on the process plant's performance and evaluate the proposed mitigation plan using an EWMA filter.

8.3.3 ARX-based Kalman Filter for Packet Dropout Estimation

In recent research, auto-regressive model with exogenous inputs (ARX) and Kalman filter have been applied for providing reliable signal estimation solutions. The t-step-ahead prediction algorithms developed using first-order ARX model and a Kalman filter to identify linear-time-invariant systems such as near future prediction of individual specific T1DM blood glucose dynamics [174] are not suitable for nonlinear system identification. Linear ARX structure can be represented by using a backward shift operator however, the parameter estimation requires transformation using transfer functions instead of a direct linear model [175]. This makes the algorithm processing more complicated. The nonlinear-ARX-type recurrent neural network (NARX-RNN) [176, 177], on the other hand, requires repetitively computational cycles of a complex neural network control system on the controller. This causes extensive delay to the control system and degrades its respective control performance. Hence, in this work, an explicit ARX-based Kalman filter is proposed for estimation of the nonlinear packet receiving rate which will be used to provide good approximation of the packet dropout rate. The approach utilizes the linear ARX model for nonlinear system identification and the advantages of the Kalman filter [178, 179] in optimizing estimation problems. The designed algorithm is benchmarked against the optimal traditional second- and third-order ARX ones.

In general, the algorithm has three steps. At first, the ARX-based state space model is formed. Second, a Kalman filter is used to estimate the state of the developed model resulting in receiving rate estimation. Finally, packet dropout rate is calculated based on this estimation.

In the first step, this work considers a packet receiving system described in (8.1) and (8.2):

$$X_k = AX_{k-1} + v_{k-1} \tag{8.1}$$

$$Y_k = BX_k + w_k \tag{8.2}$$

where:

X_k is packet receiving state.

Y_k is packet receiving rate (systems output).

X_{k-1} is past packet receiving state.

v_{k-1} is state covariance.

w_k is measurement covariance.

A is state coefficient matrix.

B is output coefficient matrix.

Based on the ARX model, the model state is set as the estimated packet receiving rate:

$$X_k = y_k \tag{8.3}$$

State space coefficient matrix is calculated as the average of the estimated coefficients:

$$A = \left[\frac{1}{N} \sum_{p=1}^{N} \hat{\beta}_p \right] \tag{8.4}$$

Hence, based on (8.1), state covariance can be found:

$$v_{k-1} = X_k - AX_{k-1} \tag{8.5}$$

The system's output is set to be real measurement of packet receiving rate:

$$Y_k = y_k \tag{8.6}$$

And the system's output gain is average weight of all the individual measurement gains:

$$B = \left[\frac{1}{M} \sum_{k=1}^{M} Y_k X_k^{-1} \right] \tag{8.7}$$

Here, the measurement covariance is found:

$$w_k = Y_k - BX_k \tag{8.8}$$

The system's dynamics is formed based on the equation set of (8.1), (8.2), (8.5) and (8.8). Therefore, a Kalman filter can be used to estimate packet receiving rate with the objective of minimizing sum squared error (SSE).

After the second step (applying Kalman filter) the estimation results are presented in Figure 8.13.

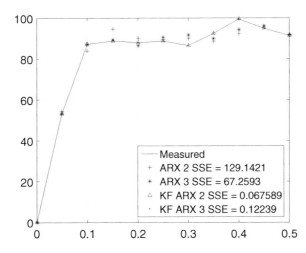

FIGURE 8.13: Packet receiving estimation using ARX 2, ARX 3, Kalman filter (KF) ARX 2, KF ARX 3 and packet receiving measurement

It can be seen that the ARX-based Kalman filter outperforms the optimized second- and third-order ARX models to achieve SSE close to 0.1. In addition, it follows well the measurement curve. The average processing time for ARX 2, ARX 3, Kalman filter (KF) ARX 2, KF ARX 3 are 0.0207 s, 0.0187 s, 0.0887 s, 0.0210 s, respectively.

Overall the ARX-based Kalman filter requires more processing time as compared with the original ARX model. It is observed that the third-order models require less processing time than the corresponding second-order models. In addition, the ARX-based Kalman filter approach results in the best estimation, thus, it is suggested to be used for estimating the packet receiving rate. It is worthy to note that except for the case of the ARX-2 based Kalman filter, most of the estimation algorithms have well below 0.05 s processing time. This is aligned with the added delay in the measurement set to be 0.05 s.

Since ARX-based Kalman filter helps to estimate packet receiving rate close to the actual measurement values at a very low SSE of 0.12239 for 10 data points, the calculated packet dropout rate is similar to the value found in Section 8.3.2.

8.4   Control Performance Assessment

This section presents the assessment of control performance of using the designed EWMA filter for addressing delay and packet dropout in industrial process plants.

8.4.1   Controller Parameters

To demonstrate the effectiveness of the proposed filtering technique, first-order and second-order process plants are chosen as they represent characteristics of many typical process plants. The considered controller delay and packet dropout are based on the findings in Section 8.3. In addition, the PID with derivative filter structure shown in (8.9), a commonly used structure for PID controllers [59], is used for regulating the process control signals.

$$C(s) = K_P + K_I \frac{1}{s} + K_D \frac{K_N}{1 + K_N \frac{1}{s}} \tag{8.9}$$

where K_P, K_I, K_D are typical PID controller parameters and K_N is a derivative filter coefficient.

The process models and their respective controller parameters, are presented in Table 8.6. The robust response time technique [180] is used to obtain the controller parameters as it enables fine-tuning the controller based on the required response time and the transient behavior.

TABLE 8.6: First- and second-order plants and their controller parameters

Plant Order	Process Plant	K_P	K_I	K_D	K_N
First-order	$P(s) = \frac{100}{s+100}$	0.036	3.376	0	100
Second-order	$P(s) = \frac{0.1013}{s^2+0.2325s}$	1.358	0.006	3.763	1.018

8.4.2 Setpoint Tracking Performance Using the Designed EWMA Filter

In this section, the setpoint tracking performance of the selected first- and second-order plants with and without packet dropout in comparison with ordinary PID, Smith predictor and PID with Kalman filter is analyzed.

8.4.2.1 First-order Plant

To assess the controller's performance, process output is studied. For a supervisory control application, control design should be 4 to 10 times faster than process time constant [181]. Based on Table 8.6, the process time constant is: $\tau_P = 1$ s. In this case, control design should be able to update process parameters within 0.1 s (10 times faster than processing time constant). Therefore, to eliminate the delay effect, (7.9) should be valid for $\forall t, t \geq 0.1$ s which means, in the worst case, (7.9) should be valid for $t = 0.1$ s. By using a graphical approach demonstrated in Section 7.2.2, similar to Figure 7.6, solution for (7.13) can be found: $x \approx 7$. From (7.14), $\tau_F = 0.0143$ s. The EWMA filter is designed based on these values. Since the process plant is sampled at period $\triangle t = 0.1$ s, from (7.3), $\alpha = 0.8749$. Hence, the EWMA compensator parameter is set. The plant responses in the presence of the measured delay are presented in Figure 8.14.

In the figure, the plant with different controller configurations has good setpoint tracking performance (at both positive and negative setpoints). The plant with original PID, Smith predictor, and PID and Kalman filter has longer settling times. That with PID and Kalman filter has the highest percent overshoot. Although the Smith predictor helps to compensate for network induced delay resulting in the least percent overshoot, its plant's response is oscillating the most. The plant with the EWMA filter has the lower overshoot as compared to that with PID and Kalman filter although it has a slight longer rise times. When the filter parameter is adjusted to $\alpha = 0.7673$ based on (7.20) to accommodate the packet dropout rate of about 12.29%. It is clearly seen that this adjustment results in further overshoot reduction in the process responses even with slightly increase in rise time. The effectiveness of this adjustment for packet dropout compensation can be observed in Figure 8.15. In terms of settling time, the plant with Smith predictor is the longest (i.e., 16.0034) while that with dpEWMA is the shortest (i.e., 15.6125 s).

To quantify the tracking performance of the plant subject to delay, Table 8.7 is outlined given the key performance parameters: integral absolute error

(IAE), rise time, settling time, and percent overshoot. The formula for the IAE is given as

$$IAE = \int |e(t)| \qquad (8.10)$$

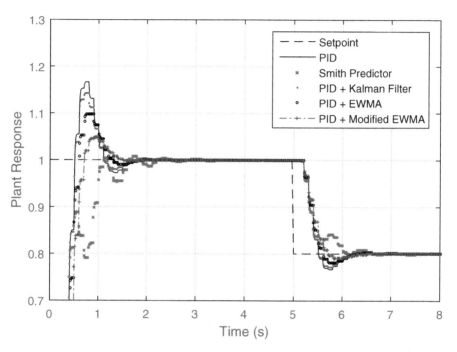

FIGURE 8.14: Responses of first-order process with delay

TABLE 8.7: Performance of first-order process with delay

Parameter	PID	Smith Predictor	PID + Kalman Filter	PID + EWMA	PID + dpEWMA
IAE	40.2486	42.9002	38.8334	40.6705	42.2504
Rise Time (s)	0.2035	0.2058	0.2043	0.2175	0.2984
Settling Time (s)	15.9286	16.0034	15.9133	15.9142	15.6125
Overshoot (%)	45.8577	26.0322	42.7756	37.1370	31.0215

Based on the table on IAE, the plant with PID and the EWMA filter have similar values while those for Smith predictor and modified EWMA filter are the highest. The plant with Kalman filter has the lowest IAE. The plant with all controller configurations has rise time of approximately 0.2 s except for the modified filter with 0.3 s. This shows that the plant with the filter has slightly slower response in setpoint changes. The settling time of the Smith predictor is the longest since it has the most oscillating response. The plant with the Smith predictor has the lowest percent overshoot. This is an indication that the Smith predictor is a delay compensator under ideal conditions. The overshoots of the plant with PID and Kalman filter are the highest, while significant overshoot reductions can be achieved by using EWMA filters (about 10% and 20% reductions for the filter and its modified one, respectively).

Taking packet dropout in a wireless plant into consideration, the plant is simulated and its responses are presented in Figure 8.15. In general, when packet dropout occurs the overshoot is significantly affected. In particular, the overshoot for the plant with delay only (below 1.2, see Figure 8.14) is lower than that with both delay and packet dropout (above 1.25, see Figure 8.15). Similar to the previous cases, overall, the plant is stable with all controller types. The plant response is similar to that with delay only. In detail, the plant with original PID exhibits the highest overshoot, followed by that with EWMA filter. Even though the the plant with the Kalman filter has lower overshoot compared to that with the EWMA filter, the modified version of the EWMA filter helps to reduce overshoot to less than that of the Kalman filter. The plant with a Smith predictor has the lowest overshoot; however, it is oscillating.

The control performance of the plant under this condition is detailed in Table 8.8. The plants with PID and EWMA filters have similar IAE (about 46.5000), while that with a Smith predictor has slightly lower value (43.8411) and that with a Kalman has the lowest value (39.1041). With regard, to rise time, the plant with the Kalman filter has the lowest value. This means the Kalman filtered plant has the fastest response to setpoint changes. Even though its settling time (15.9132 s) is still slightly longer than that of the modified EWMA filter (15.6061 s). Because of the fastest response, the Kalman filtered plant has moderately high percent overshoot, at 42.7784%, which is 16% lower than that of original PID. The plant with the modified EWMA filter has lower percent overshoot, at 37.8157%, while that with the Smith predictor has the lowest value of close to 25%.

Overall, the major advantage of EWMA is that it provides better control performance by a significant reduction in percent overshoot as compared to other types of controllers except the Smith predictor. Although the Smith predictor is dedicated to delay compensation, its performance is the worst as the plant response is most oscillating.

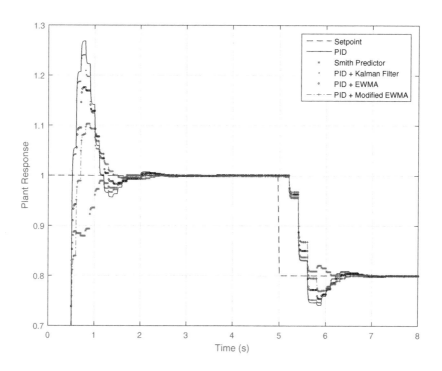

FIGURE 8.15: Responses of first-order process with delay and packet dropout

TABLE 8.8: Performance of first-order process with delay and packet dropout

Parameter	PID	Smith Predictor	PID + Kalman Filter	PID + EWMA	PID + dpEWMA
IAE	46.8435	43.8411	39.1041	46.4197	46.5848
Rise Time (s)	0.2997	0.3021	0.2043	0.3016	0.3065
Settling Time (s)	16.0003	16.0054	15.9132	16.0018	15.6061
Overshoot (%)	58.4984	25.3440	42.7784	46.9556	37.8157

8.4.2.2 Second-order Plant

With the controller parameters described in Table 8.6, the process time constant is $\tau_P = 4.17$ s which is longer than the process time constant of the first-order plant in Section 8.4.2.1. Thus the EWMA filter parameter remains unchanged. For setpoint performance assessment, the plant responses are presented in Figure 8.16. It is seen from the figure that the plant controllers are stable. In addition, that with PID and Kalman filter has similar responses with the highest overshoots as they have the shortest rise times. On the other hand, although response speed is slower, the plant with the Smith predictor has lower overshoot. Its magnitude is slightly lower than that with the EWMA filter. However, when the filter is modified, it can result in the lowest overshoot as indicated in the figure. As a drawback, its response to setpoint change is also the slowest.

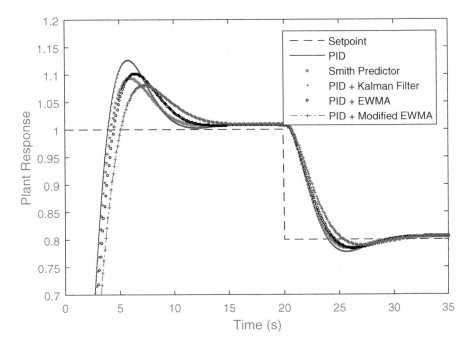

FIGURE 8.16: Second-order process with delay responses

The detail control performance of the plant is presented in Table 8.9. As seen from the table, the plant with all controllers has IAE values about 40 to 50. Although the EWMA filter and its modified filter have the highest values and longer rise times, their settling times are among the shortest. In addition, their percent overshoots are also about 2% to 4% lower than that of the plant with original PID and the Kalman filter, respectively. The plant that has the second lowest percent overshoot is that with Smith predictor, at 8.7006%, which is still about 1.5% higher than that of the modified filter.

TABLE 8.9: Performance of second-order process with delay

Parameter	PID	Smith Predictor	PID + Kalman Filter	PID + EWMA	PID + dpEWMA
IAE	43.6150	43.1454	43.8500	46.0657	48.8168
Rise Time (s)	2.5418	2.7122	2.5630	2.8951	3.3060
Settling Time (s)	46.7755	43.6033	46.8344	43.8384	44.3064
Overshoot (%)	11.9527	8.7006	11.9453	9.3971	7.2318

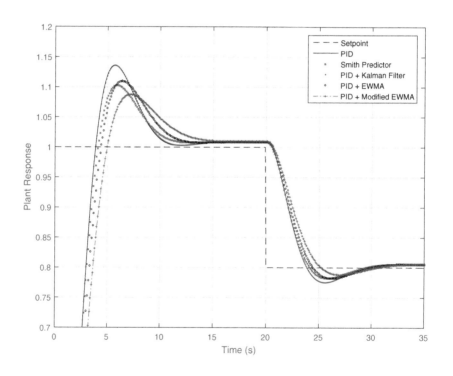

FIGURE 8.17: Second-order process with delay and packet dropout responses

The performance of the plant when both delay and packet dropout are taken into consideration is presented in Figure 8.17. Based on the figure, when packet dropout occurs, the plant with all controllers has a slight increase in peak percent overshoot. For example, at peak, the plant with PID has overshoot increases from 1.125 to 1.128. Although there is increase in overshoot, the plant is stable for all configurations of the controllers. The response of the plant in this case is similar to that in Figure 8.16.

The detail performance of the plant under this condition is outlined in Table 8.10. Overall, the plant performance is similar to that obtained in Table 8.9 with a slight increase in all control parameters. Still, the plant with EWMA filters is the best in reducing percent overshoot. Thus, this is the key advantage of using the EWMA filter in a process plant.

TABLE 8.10: Performance of second-order process with delay and packet dropout

Parameter	PID	Smith Predictor	PID + Kalman Filter	PID + EWMA	PID + dpEWMA
IAE	43.7874	43.2767	44.0178	46.1653	48.8319
Rise Time (s)	2.4996	2.6621	2.5210	2.8453	3.2490
Settling Time (s)	46.8996	43.5752	46.9580	43.8096	44.2713
Overshoot (%)	12.9168	9.5811	12.8697	10.1745	7.8561

In summary the proposed filter is not only effective for the first-order system, but also for a more general system, second-order system. In process control, the higher-order systems are usually approximated to first- and second-order systems for simplification purposes. Therefore, the technique has promising effectiveness in process plants as it does not require significant change in systems' infrastructures.

8.4.3   Disturbance Rejection Performance Using the Designed EWMA Filter

Apart from setpoint tracking performance, disturbance rejection is also an important control performance index that should be analyzed for any process plant. This section details the analysis of the results obtained for the first-order and second-order plants. When the plant is stable after setpoint change, both positive and negative disturbance with magnitude of 0.1 (10% of original setpoint) will be introduced consequently into the plants' inputs. The responses of the plants are then captured and presented.

8.4.3.1   First-order Plant

For the first-order plant, the responses to different controller settings are presented in Figure 8.18.

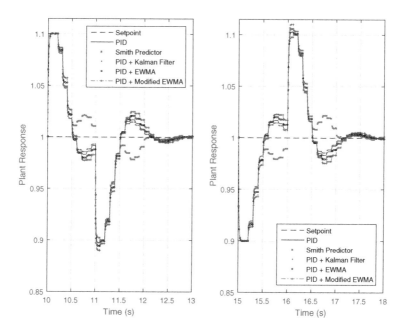

FIGURE 8.18: First-order plant responses in the presence of (left) positive, and (right) negative disturbances

Based on Figure 8.18, when disturbance is introduced, the response of the plant with the Smith predictor oscillates the most. In addition, its transition when approaching the setpoint (i.e., between 10.5 s and 11 s, moving upward) is in inverse direction compared to those of other control strategies (i.e., moving downward). The plant with the EWMA filter has slightly slower response to the disturbances while that with the Kalman filter is faster as compared to the plant with PID only. Because of the slower response to disturbance, EWMA filters introduce slightly increased overshoot. However, when considering the case of negative disturbance, it helps to maintain the past response longer. Thus, in case of packet dropout, this helps to maintain the plant response around the good setpoint before a new good control command is received.

Further detail on control performance of the plant is presented in Table 8.11. Based on the table, the plant with EWMA filter and Smith predictor has the highest IAE values, while that with Kalman filter has the lowest value. Since the Smith predictor has the longest rise time (0.7 s) compared to other controllers ($\approx$0.2 s), its response is oscillating. Because of a slightly longer rise time, the settling times of the plant with EWMA filters are slightly longer than those of PID and PID with Kalman filter. Still in terms of overall overshoot, the plant with EWMA filters has the least values. This is because the filters were used in partial configuration during the setpoint change.

TABLE 8.11: Performance of first-order process with disturbance

Parameter	PID	Smith Predictor	PID + Kalman Filter	PID + EWMA	PID + dpEWMA
IAE	36.3480	38.5137	35.1601	37.2382	39.0232
Rise Time (s)	0.2972	0.7174	0.2988	0.3085	0.3978
Settling Time (s)	16.4212	16.8188	16.4220	16.8043	16.9090
Overshoot (%)	16.6868	10.0826	14.2207	10.5976	11.0230

8.4.3.2 Second-order Plant

When disturbance occurs, the second-order plant responses are presented in Figure 8.19. Similar to the case of the first-order plant, the response of EWMA filters for the second-order plant is slower to the introduced disturbance as compared to that with PID only, Kalman filter and Smith predictor. This results in slightly increased overshoots due to the disturbances. However, after reaching the peaks due to the disturbances, the plant with the EWMA filter has response close to that with PID and Kalman filter. While that with the modified filter has response close to that with the Smith predictor. When negative disturbance is introduced, the responses of the plant with all controllers are similar at first. After reaching the minimum point, the responses of that with PID, Kalman filter and EWMA filter are similar, while those with modified EWMA filter and Smith predictor are similar.

The performance of the plant is further detailed in Table 8.12. Similar to the case of first-order plant, the second-order plant with EWMA filters has slightly higher IAE, longer rise time and settling time, while its percent overshoot is significantly lower than that of the plant with PID, Smith predictor and especially Kalman filter.

Despite the 10% input disturbance, the plant with the modified filter has the lowest percent overshoot, at 7.4834% which is the result of partial configuration during the setpoint change.

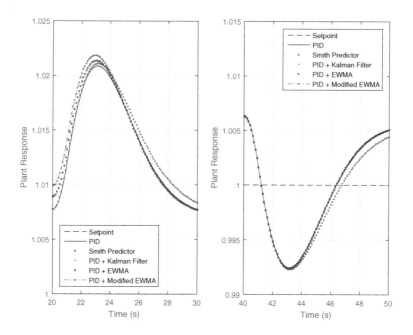

FIGURE 8.19: Second-order plant responses in the presence of (left) positive, and (right) negative disturbances

TABLE 8.12: Performance of second-order process with disturbance

Parameter	PID	Smith Predictor	PID + Kalman Filter	PID + EWMA	PID + dpEWMA
IAE	34.4930	34.2771	34.6710	36.2874	38.2944
Rise Time (s)	2.5401	2.7100	2.5613	2.8887	3.2926
Settling Time (s)	9.2788	9.5663	9.3920	10.3874	11.6445
Overshoot (%)	12.0050	8.7538	11.9975	9.5491	7.4834

In summary, it is seen that the key advantage of the plants with EWMA filters is the significant reduction in percent overshoot as compared to those with original PID, Smith predictor and Kalman filter.

8.4.4 Simulation Time Comparison with Smith Predictor and Kalman Filter

This section performs the comparison in terms of simulation time between the proposed EWMA filter, Smith predictor and Kalman filter. The comparison takes into consideration the change in process dynamics, in particular, its transfer function's numerator. The selected plant model is of second-order as it represents better system dynamics than many process plants. The process sampling period is 0.1 s with end time of 80 s and the simulation experiment is carried out using a Samsung laptop running Windows 7 x64 with Intel Core i7 2630QM CPU at 2.0 GHz and 4.0 GB RAM. As a result, the average simulation time for 100 trials each category is presented in Figure 8.20.

It is seen from the figure that the simulation time for the plant with the Smith predictor is the most stable despite the changes in the plant's transfer function's numerator. Without process mismatch, i.e., at numerator, the overall simulation time of the Smith predictor is lower than that with model mismatch. The simulation time when the plant has a Kalman filter varies the most and is the longest as well, standing at about 0.75 s which is about twice the simulation time of the plant with either EWMA filter or Smith predictor. This is because of the iterative nature in the operation of a Kalman filter. The simulation time of the plant with EWMA filter is also stable around 0.35 s which is slightly higher than that of Smith predictor at 0.34 s.

The detail standard deviation of the simulation time for each case is presented in Table 8.13. From this table, the process with Smith predictor has the lowest standard deviation, followed the process with EWMA filter and Kalman filter. Overall, the Smith predictor is more sensitive when process parameter deviates from its nominal value; the Kalman filter is more complex and thus results in significantly much more processing time; EWMA filter has average simulation time and is less susceptible to process parameter changes.

TABLE 8.13: Average simulation time deviation

Mismatch	Smith Predictor	Kalman Filter	EWMA Filter
No Mismatch	0.010	0.014	0.018
+5%	0.014	0.016	0.015
+10%	0.013	0.013	0.009
-5%	0.008	0.015	0.013
-10%	0.012	0.022	0.017

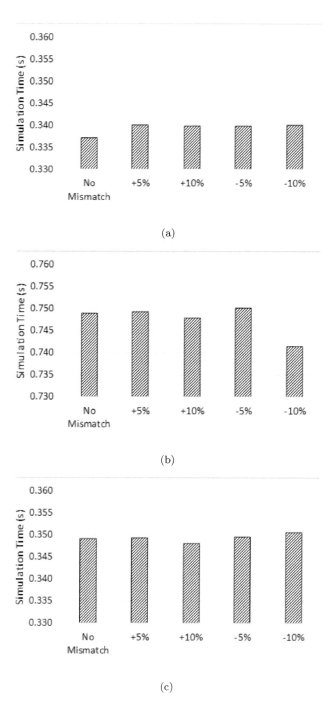

FIGURE 8.20: Average process simulation time: (a) Smith predictor, (b) Kalman filter, and (c) EWMA filter

8.5 Summary

In this chapter, it can be seen that wired link contention contributes to delay and packet dropout to the overall wireless networked control system. It can be effectively addressed by using the advanced EWMA filter design technique presented in Chapter 7. The performance of the filter is detailed and proved via experiment with a WirelessHART evaluation kit. In the next chapter, the WirelessHART hardware-in-the-loop simulator will be presented. It serves as a validation platform for any wireless control algorithms since actual hardware are involved in the real-time wireless communication. The filter's performance on wireless links of an industrial-scaled laboratory environment will be presented in the following chapter.

9

Wireless Hardware-in-the-loop Simulator

9.1 Introduction

One of the key challenges for applying any novel control techniques to an industrial process plant is the actual proof-of-concept using the algorithms. It is well known that not all process industries are available for testing and trials of the newly developed algorithms. Thus, hardware-in-the-loop simulation is a potential approach to close the gap between academic and industrial applications. In this chapter, a novel WirelessHART hardware-in-the-loop simulator will be presented. Its architecture will be detailed to assist readers in applying the approach for their own applications.

9.2 WirelessHART Hardware-in-the-loop Simulator (WH-HILS)

9.2.1 WH-HILS Architecture

The architecture of WH-HILS is presented in Figure 9.1. In the figure, the simulator essentially consists of a WirelessHART gateway (LTP5903CEN-WHR), several wireless nodes (DC9003-A), and a computer. The gateway with an embedded access point serves as the network and security manager, and as the source of network information. This gateway communicates with other nodes in burst mode in which, if a device is busy, the received message will be placed in queue and the communication can be continued. The nodes function as wireless field devices installed at process plants. In addition, each of them can serve as a relay for routing messages to other nodes in a multi-hop network. The controller and process plant functionalities can be developed and integrated into the computer using MATLAB®. By using Real-time Sync block in Simulink, the simulation is synchronized with the computer's clock, thus it can be run in real-time. For interfacing with the gateway, a Python program is used. This interface is natively supported by MATLAB®2015 and later versions and is described in detail in the following section. Currently, the supported Python versions are 2.7, 3.3, and 3.4 [182].

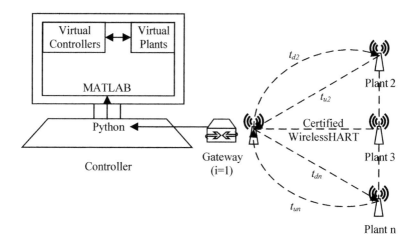

FIGURE 9.1: Real-time WirelessHART hardware-in-the-loop simulator (WH-HILS)

9.2.2   WH-HILS Device Interface

For device interface, the selection of communication protocol is important for ensuring reliable communication between devices. This section describes and justifies the selection of communication medium mainly between the computer and the gateway as shown in Figure 9.2. This is because the node can operate as a standalone device as mentioned in the previous section. Although the WirelessHART gateway supports three protocols for communication with computer, namely serials 1 and 2, and local area network (LAN) cable. LAN interface is selected as it has the highest reliability. In the figure, the physical (PHY) layer between the gateway and the computer is the LAN interface. The gateway has built-in application (APP) layer with extensible markup language remote procedure call (XML-RPC) server which can supply all network information upon request from the controller at the computer.

The first section of the APP layer consists of several Python functions written specifically to communicate with and retrieve information from the gateway server. The second section of the APP layer is the built-in MATLAB Python library which allows MATLAB to call and run external Python functions within its own operating environment. The information passing through this section will be automatically converted to corresponding data type supported by MATLAB or Python, respectively. The top section of the APP layer is the working environment of MATLAB where various controller and process

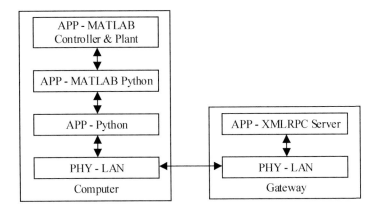

FIGURE 9.2: WH-HILS computer-gateway communication stacks

plant models can be developed. The simulation can be run using Simulink blocks and/or commands.

Initial configuration for each node can be done through serial interface with a computer. Once configured in master mode, after powering up, the nodes will automatically search for and connect to the gateway's network. They can self-operate with built-in batteries. The maximum distances between the nodes and the gateway are determined by several factors including wireless power transmission, receiver's sensitivity of the devices, and surrounding obstacles. The pilot process plant laboratory environment is considered in this research.

9.2.3 WH-HILS' Potential Industrial Applications

The developed WH-HILS requires only a computer, wireless nodes and a gateway, thus it can be considered a mobile platform for investigation of WHNCS for industrial process control and monitoring applications. This is useful and cost saving in the following two scenarios. Scenario 1: a wired plant is planning to upgrade its facilities to wireless. The assessment can be done by mobilizing the system to the plant site, installing wireless nodes and gateway to designated locations, measuring real-time delays and assessing control performance of the process plant virtually. Because the gateway is certified, it can communicate with any industrial WirelessHART transmitters for flow rate, pressure, level measurements. These transmitters can replace the nodes if required. Scenario 2: a wireless plant is planning to adopt a new control algorithm. A feasibility study on the effectiveness and reliability of the control algorithm can be performed at the plant site without interfering in its operations. This leads to significant cost saving for adoption and deployment of the new technology into the plant since any operation disruption or shutdown results in severe economic loss to the host company.

9.3   Real-time Delay Measurement Experiment with WH-HILS

This section details the procedure for real-time round trip delay measurement using the WH-HILS. As seen from Figure 9.1, for any node i (i = 2, 3,..., n), the upstream delay (t_{ui}) is the delay when sending signals from the node to the gateway and the downstream delay (t_{di}) is the delay when sending signals from the gateway to the node. These two types of delays can be measured by the gateway using the command *getLatency MACaddress* where *MACaddress* is the media access control (MAC) address of the nodes connected to the gateway [183]. Similar to [4], time delay is measured using timestamps on communication messages. In particular, the measured delay is based on the difference between the received timestamp generated at the gateway and the sent timestamp embedded in the arrival message. Although each device in the network has its own internal clock (e.g., quartz crystal) and the clock experiences time drift during operation [19], network-wide time synchronization prescribed by the standard can ensure all devices have the same time reference as the gateway with better accuracy of less than 1 ms [97]. This makes the delay measurement from gateway feasible and accurate.

The experimental setup follows the WH-HILS architecture of Figure 9.1 with one access point and five nodes (i.e., n = 6) deployed in a pilot process plant laboratory shown in Figure 9.3.

The measurement procedure is detailed in Figure 9.4. Initially, the nodes are turned on sequentially from top to bottom of Figure 9.3 to join the network formed by the gateway. Each node is assigned a short identification number (ID) upon successfully joining the network. The access point always has the ID of 1 since it is the first to establish connection with the gateway. In order to make sure the nodes have the IDs sequentially from 2 to 6 as shown in Figure 9.3, each node was turned on and it was ensured to successfully join the network before the next one is turned on. In the case of network formation with multiple nodes turning on at the same time, the assignment of IDs for the nodes depends on the automatic searching and negotiation processes between the nodes and the gateway. In that case, the ID assignment is based on the first-join-first-get basis so the order of the nodes' IDs is random. Then, an initialization program is launched to establish connection between the gateway and the controller. Next is the initialization of logging variables for capturing the delays from each node. Lastly, the WH-HILS starts to record the delays for a specific simulation duration.

In order to emulate external interference from operation of similar indus-

FIGURE 9.3: Pilot process plant laboratory layout with node IDs. 1: Computer, WirelessHART gateway and access point (GW & AP), 2: Level pilot process plant (LV), 3: Heat exchanger process pilot plant (HE), 4: Flow process pilot plant (F), 5: Boiler drum & heat exchanger process pilot plant (BL & HE), 6: Air flow, pressure and temperature process pilot plant (AF, P, T)

trial plants, three hours after the start of the experiment, the pilot process plants collocated with nodes 2, 3 and 4 will be turned on and they will be run for the next three hours.

9.4 Summary

In summary, this chapter has presented a novel WH-HILS platform for validation of wireless control algorithms. The simulator is suitable for use with any control algorithms and it is scalable. This means, the simulator can be expanded for a more complicated network with more wireless nodes. This provides a great flexibility in both academic and industrial use cases.

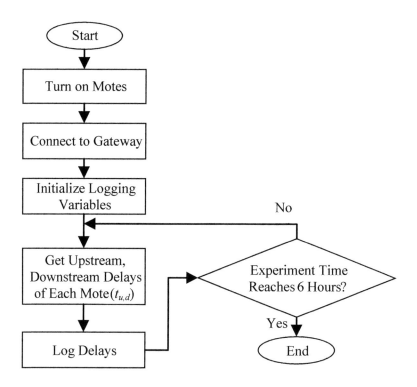

FIGURE 9.4: Delay measurement procedures with WH-HILS

10

The Filter's Performance over Wireless Links of Industrial-scaled Wireless Mesh Networks

10.1 Introduction

This chapter presents the experiment results using the developed real-time WirelessHART hardware-in-the-loop simulator (WH-HILS) where actual wireless gateway and five nodes are used. The process plants used in the experiment are industrial-rated, thus the surrounding environment is close to industrial environment, except that the laboratory is maintained at a temperature of around 22°C according to the university's laboratory rules. This is to extend the lifetime of the laboratory equipment for educational purposes.

10.2 Results with Real-time WirelessHART Hardware-in-the-loop Simulator (WH-HILS)

10.2.1 Process Plant Model Selection

As discussed in Section 9.2.1, the simulator is able to simulate any type of process plants and controller for demonstration purposes, and the selected plant models described in Table 8.6 will be used for validation with the developed WH-HILS. Both upstream and downstream delays are taken into consideration for tuning the plants' controllers. Since the wireless delay is time-varying, the average value downstream delay of 0.320 s presented in Section 9.3 is used for tuning the process plants. This is because the delay is consistent since the gateway will manage to schedule communication to the motes evenly. In addition, the delay is small enough compared to the delays reported in Table 8.2 (1.280 s and 1.626 s for downstream and upstream delays, respectively). With the low-delay setup, the total round trip average delay is below 1 s, hence the controllers are re-tuned to achieve 2 s response time which represents well dynamics of fast processes. As a result, their controller parameters are presented in Table 10.1. As seen from the table, the controller for the first-order plant is an integral controller, while that for second-order plant is a complete PID controller.

TABLE 10.1: First- and second-order process plants with controller parameters

Plant Order	Process Plant	K_P	K_I	K_D	K_N
First-order	$P(s) = \frac{100}{s+100}$	0	1.0	0	100
Second-order	$P(s) = \frac{0.1013}{s^2+0.2325s}$	5.1272	0.0442	8.0229	5.4540

10.2.2   Experiment Overview and Network Configuration

The experiment was conducted using the developed simulator with the run-time of six hours. It is used to assess the control performance of the proposed filter in comparison with other common control approaches using PID, Smith predictor and Kalman filter in the presence of time-varying mesh network delays. The network consists of five nodes, and the delays of the first two nodes, 2 and 3, are used for assessment of the first- and second-order plants. As mentioned in Section 9.3 three hours after the start of the experiment, the pilot process plants collocated with nodes 2, 3 and 4 will be turned on and they will be run for the next three hours. This is to emulate external interference from operation of similar industrial plants.

The default service level agreements (SLA) in the gateway are set to achieve: minimum 99.00% reliability, maximum network delay of 125 s, and minimum network path stability of 50.00%. The traffic generated by these nodes is set to 2 packets per second, each with a length of 10 bytes (more than sufficient to report a process parameter) by using PkGen software provided within the SDK from Linear Technology. With this setting, the average traffic generated by nodes 2, 3, 4, and 6 are 2 packets per second while that by node 5 is 3 packets per second.

As defined by the standard, the superframe is a set of time slots which is assigned for communication of a pair of devices, thus it is repeatable, periodical and configurable. The minimum default bandwidth profile (P1) has 1024, 256, and 128 time slots for upstream, downstream, and advertisement superframes which is suitable for industrial monitoring applications [97]. By further reducing these minimum values, the nodes will be able to communicate with the gateway more often, thus reducing network delay. The above fast traffic is achievable by setting the bandwidth profile to have shorter superframe lengths of maximum 256, 64, and 32 time slots, respectively. This profile is used for demonstration of a low-delay WirelessHART network.

10.2.3   Network Statistics

Basically, the network device statistics (obtained through Command 840) is represented by its reliability, average upstream delay and stability as shown in Table 10.2.

TABLE 10.2: Network statistics overview

Item	0 h	2 h	4 h	6 h
PkLost	1	1	1	1
Total PkTx	243	95k	206k	330k
(PkFail)	49	25k	53k	93k
Total PkRx	3165	76k	161k	243k
Reliability (%)	99.97	100	100	100
Average Latency (s)	3.07	1.12	1.05	1.07
Stability (%)	79.84	73.11	74.24	71.81

Note: k = 1,000.

It is seen from the table that the network's reliability is almost 100%. During the entire 6-hour period of the experiment, only one packet was lost in the network. This means almost all packets sent in the network reach the desired destinations. This high reliability makes WirelessHART robust for industrial monitoring application. On average, the network's stability given by $PkFail/TotalPkTx * 100\%$ is above 70%. Here, $PkFail$ is the number of packets that failed during the transmission, while $TotalPkTx$ is the number of packets that have been sent (including the retransmissions of failed packets) by network devices. At 0 h, the round trip delay is 3.07 s. However, in the later hours, the delay falls to approximately 1 s. This was achieved through setting each mote to generate 2 packets per second on average. On an hourly basis, health report (report device health - Command 779, read neighbor health list - Command 780) is captured by the gateway and used by the manager in optimization and diagnostics. Using diagnostic information, the gateway is able to self-optimize by improving each individual path stability, hence maintaining overall network stability to above 70%. The self-optimization functionality is another important adaptability feature of WirelessHART which makes it suitable for industrial applications. Referring to the table, it is clearly seen that the operating process plants (plants 2, 3, and 4) during the second half of the experiment (from 3 h to 6 h) result in a significant increase in packet fail rate from 25k for the first two hours to 40k for the last two hours. However, here, the packet fail concept should not be misunderstood with packet loss. This is because a failed packet can be retransmitted. Still the stability is above 70%. On average, the network delay is reduced from 3.07 s to 1.12 s, then 1.05 s for the first four hours and slightly increases to 1.07 s at the end of the experiment. This is because of the increase in packet fail rate during the last two hours, thus, resulting in the increases in packet regeneration rate, network traffic, and delay.

10.2.4   Self-Adaptation Network Topology

By reading network topology and actually implemented links' information (Command 834), the changes in network topology are obtained and presented in Figure 10.1. At the beginning of the experiment, i.e., 0 h, the network topology formed is a mesh network (see Figure 10.1a). Here, the access point communicates directly to nodes 2, 3, 5, and 6. Node 2 links with nodes 4 and 5, while node 4 links with nodes 3 and 6. After two hours, at 2 h, the mesh network has been changed to a more complex one with an additional one direct link between the gateway and node 4. Since then, the network topology did not change. It was maintained even after plants 2, 3, and 4 were turned on at 3 h (Figure 10.1b). To allow for sufficient time to observe the network statistics, the plants were allowed to run for another three hours.

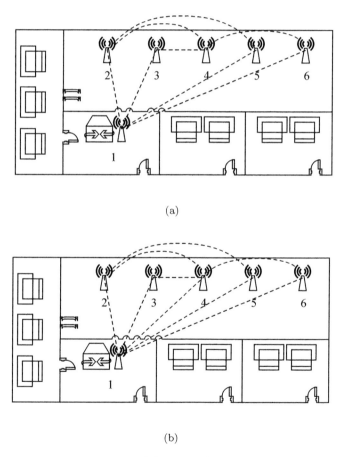

(a)

(b)

FIGURE 10.1: Network topology (a) at 0 h, (b) at 2 h, 4 h and 6 h. Note: the plants 2, 3, and 4 were turned on at 3 h.

10.2.5 Network Background RSSI-Stability Curve

In order to assess the network's characteristics, an important measure that should be investigated is the average background RSSI and path stability curve (known as the waterfall curve). To facilitate this analysis, the curve is presented in Figure 10.2 based on 24 reported 15-minute network statistics spanning the 6-hour experiment. It should be noted that here, the background RSSI's definition is different from the conventional definition of RSSI and it represents the index of a wireless device's background noise, computed as in [97]. The higher the background RSSI (measured in dBm) the higher the receiver's background noise and the weaker the received signal.

With regard to the SLA preset in Section 10.2.2, the network's performance is acceptable as both plots in Figure 10.2 show that the overall network stability is above 50% and the majority of the links have background RSSI below -60 dBm. On average, the direct node-gateway paths have background RSSI in the range from -80 dBm to -70 dBm while the node-node paths have that in the range from -70 dBm to -64 dBm. As observed from Figure 9.3, the gateway is placed at a higher location and is clear of any obstruction. On the other hand, all nodes are placed on top of the control panels which are made of metals and surrounded by cables and other metal frames which can be considered as light obstruction. This obstruction results in higher background RSSI for the nodes, hence, more noise or interference exists around them.

The key difference between Figures 10.2a and 10.2b is at the scatter plots. Before the plants were turned on, the average background RSSI-stability is more condensed (Figure 10.2a). After the plants were turned on and running, the average background RSSI-stability is more scattered (Figure 10.2b). This is observable at both node-access point and node-node paths. In addition, together with the spread on the second plot, it is shifted slightly to the right (of x-axis) and the bottom (of y-axis). Hence, it is evident that the operating plants introduced more noise (more paths having average background RSSI above -70 dBm; Figure 10.2b) to the network and this leads to reduced stability. Furthermore, the node-node paths are affected the most because all nodes are collocated near the operating plants. This is observed by comparing the stability histograms (on right panes) of the two plots in Figure 10.2.

In summary, when the process plants are in operation, the wireless communication of the network is affected mostly at the communication between the surrounding nodes. This is due to the additional background noise generated by the operating equipment i.e., pump, heat exchanger, and other electronic equipment. In addition, the self-organizing and self-healing characteristics of WirelessHART were demonstrated through the change in mesh topology during the experiment.

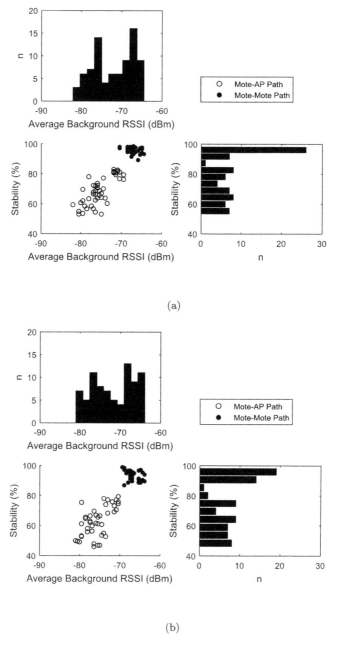

(a)

(b)

FIGURE 10.2: Average background RSSI-stability waterfall curve, histograms of the stability, and the average background RSSI: (a) before, (b) after turning on plants 2, 3 and 4 at 3 h.

10.2.6 Measured Real-time Network Induced Delays

10.2.6.1 Gateway's Upstream and Downstream Delay Characteristics

The experiment with the WH-HILS was conducted over a period of six hours for assessing plants' control performance using real-time measured delays. Here, the downstream delay from the gateway is constant at 0.320 s which is half of the defined downstream length of 64 time slots as mentioned in Section 10.2.2. The manufacturer provides this value because if the number of downstream transmissions is very large and random, the average downstream delay is equal to the mean of the sum of all downstream delays which is half of the downstream length.

During nodes' join or rejoin events (network formation or reformation), average downstream delay is equal to the advertisement period which is set as 32 time slots in the gateway's configuration. Differently, the upstream delay varies over time since it depends on the communication between the node and the gateway. In addition, being the host in the network, the gateway is always in active state. On the other hand, the nodes will enter idle state after completing communication (in active state) with other nodes or the gateway. This is to reduce power consumption, and thus save battery lifetime of the nodes. In this idle state, the nodes will have lower processing capability, hence resulting in longer delay to process signals. This contributes to the variation of upstream delays to the gateway.

It should be noted that the gateway connected to the computer is capable of measuring exactly the communication delays to and from any nodes in the network through timestamps on the delivered messages. The overall upstream delay is expected to be mainly within the defined upstream superframe of 256 time slots or equivalently 2.560 s.

10.2.6.2 Measured Upstream Delays

The recorded upstream delays of five nodes and their corresponding boxplots are presented in Figures 10.3 and 10.4, respectively.

In Figure 10.3, considering the average delay, nodes 2, 3 and 4 have the lowest values of about 0.3 s, node 6 has the average value of 0.5 s, and node 5 has the highest value of about 3 s. In addition, for all nodes, the variation of upstream delays observably increased from 3 h toward the end of the experiment. The peak delays occurred between the 4 h and 5 h for all the nodes. This corresponds to the duration for which the plants 2, 3 and 4 were running.

From Figure 10.4, the average delay for all the nodes is below 3 s. Particularly, the value is below 0.5 s for the majority of the nodes (nodes 2, 3, 4, and 6), and is between 2 s - 3 s for node 5. In addition, node 5 recorded the median delay of 3 s with the maximum value of about 8.2 s, thus, having the highest variation.

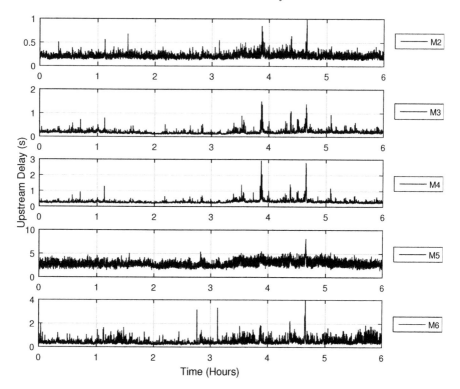

FIGURE 10.3: Measured network induced delays a 6-hour experiment with 43,201 measurements sampled in real-time, two each second

Furthermore, in Figure 10.4, all of the nodes have upper extremes (outliers) but not lower extremes. This indicates that these nodes potentially experience packet dropout in the communication which can affect performance of wireless control systems.

10.2.6.3 Suitability of WirelessHART for Fast Process Control

From the above results, it is seen that the network delay is typically in the range of 0.2 to 0.5 s, thus less than a second which is appropriate for industrial control of processes with long dead-time. Such processes are for level, pressure, and temperature control. It should be noted that the suitability of the application for control depends mainly on process response time (time constant plus dead-time).

A rule of thumb is that the control action should be executed 4-10 times faster than the response time of the particular process [181]. For example, if the wireless delay is 0.5 s, it is suitable for level process control with process response time of 5 s or more. Some typical industrial processes with time constants that are suitable for adoption of WirelessHART technology are gas flow, furnace pressure, compressor surge control (5 s), vessel pressure (30 s),

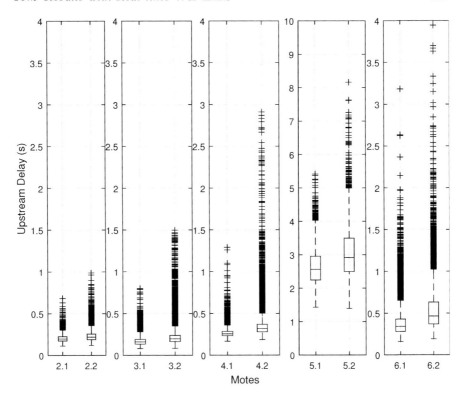

FIGURE 10.4: Boxplot of the measured upstream delays for first and second halves of the experiment, corresponding to on and off period of plants 2, 3, and 4

column pressure (50 s), gas composition - O_2 (60 s), exchanger temperature, boiler stream temperature (180 s), liquid level (300 s), vessel temperature (600 s) [1]. In these processes, the dead-time is about 10% of the corresponding time constant. Although there is transmission time variation in a wireless network, with very high reliability (up to 99.999%) supported by TSCH and suitable selection of the process to be controlled, WirelessHART is definitely suitable for wireless control in industrial applications.

For applications that require much faster update rate ($\geqslant$1000 Hz or delay less than 1 ms, e.g., motor drive, vibration monitoring), WirelessHART is still suitable. This is achievable given that its maximum speed of 250 kbps and fast-pipe communication are enabled and the communication message is less than 128 bits. At this configuration, the minimum delay is 0.512 ms ideally. By further reducing the message payload length, less than 0.512 ms communication delay can be achieved, thus it is suitable for this type of application. However, an issue with such network configuration is that it will drain the battery of the wireless node much faster. Therefore, the wireless node needs to be powered by a constant, reliable power line. Another issue is that the fast-pipe communication is only supported for a single node, therefore, only the simplest single-hop-single-path network topology can be deployed.

10.2.7   Control Performance Assessment

This section presents the analysis of control performance of the first and second-order plants using the developed WH-HILS hybrid simulator. The analysis takes into consideration both ideal and unideal plants (with model mismatch and input noise). For further highlighting the advantages of the proposed dpEWMA filters, the control performance of the plant when using the filters is compared with Everett's [184] and Tseng's [142] methods. For Everett's approach, the EWMA filter is designed based on an MA filter such that the exponentially smoothed average is the same as the average age of the sample. In this work, the average age of the sample is 0.1 s as mentioned in Section 7.2.

10.2.7.1   First-order Plant

The plant responses of the first-order process are presented in Figure 10.5.

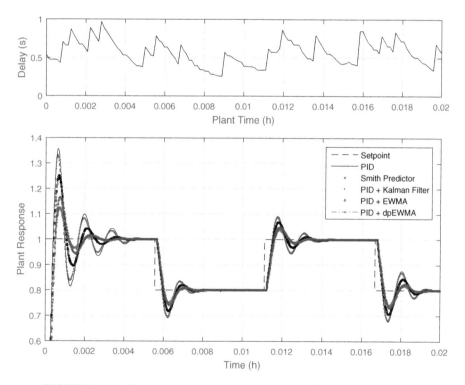

FIGURE 10.5: First-order plant responses in validation experiment

From the figure, it is seen that the response of the plant with the Kalman filter is very close to that with PID only. While the plant with EWMA filters significantly reduces the overshoot by approximately 50%. Although the plant with the Smith predictor has the best performance, its response to the setpoint change is the slowest. Overall, the plant with all controller configurations is stable. Slight oscillation is observed for the plant with EWMA filters while that has higher magnitude for the plant with only PID and with the Kalman filter.

More detail on control performance of the plant can be found in Table 10.3. From the table, it is observed that despite the existence of both time-varying upstream and downstream delay in the plant communication, the Smith predictor is able to produce the best control performance as compared to other controller types. However, its response is slower since the rise time is longer compared to that of the plant with PID only, Kalman filter and EWMA filters. Although, the EWMA filters' performances are not the best, they are still better than the original PID and that with Kalman filter.

TABLE 10.3: Performance validation of first-order process

Parameter	PID	PID + Smith	PID + Kalman	PID + EWMA	PID + dpEWMA
IAE	201.1738	146.8790	190.1970	174.0369	162.0609
Rise Time (s)	0.6838	0.8176	0.6869	0.7726	0.8032
Settling Time (s)	68.04	65.88	67.68	66.24	66.24
Overshoot (%)	70.0782	40.3809	66.6682	56.4847	45.5845

When noise and model mismatch are taken into consideration, the responses of the plant with different control strategies' are presented in Figure 10.6. From the figure, it is seen that the responses' overshoots significantly increase when there is noise and model mismatch. Although, it takes longer for the plant to settle, overall, the plant is still stable. With the dpEWMA filter, the plant has the best response compared to the responses when the Smith predictor, EWMA filter, Kalman filter and original PID are used.

More detail of the plant performance with different control strategies is presented in Table 10.4. Here, it is realized that due to the presence of noise and delay mismatch, the majority of performance factors are increased except rise time. In detail, the IAE is increased the most (about 40%) with the Smith predictor, and the least with the dpEWMA filter ($\approx$15%). While the rise time in general slightly decreases. Because of the faster rise time, the settling time is increased. Therefore, the percent overshoot increases.

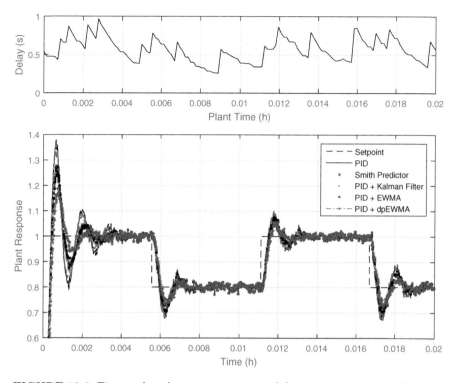

FIGURE 10.6: First-order plant responses in validation experiment with presence of noise and model mismatch

TABLE 10.4: Performance validation of first-order process with presence of noise and model mismatch

Parameter	PID	PID + Smith	PID + Kalman	PID + EWMA	PID + dpEWMA
IAE	214.0474	186.1952	202.8927	187.8248	177.1465
Rise Time (s)	0.6894	0.7214	0.6903	0.7864	0.8852
Settling Time (s)	71.64	71.64	71.64	71.64	71.64
Overshoot (%)	73.5912	60.4231	70.1995	59.9511	48.8171

Similar to the case of IAE, the Smith predictor is affected the most as the percent overshoot increases 20% in magnitude. This explains why the Smith predictor is relatively good only for ideal plants. It should be used together with other compensation approaches to have a better delay and packet dropout mitigation effect.

For more comparison with recent related approaches using Everett and Tseng's methods, Figure 10.7 presents the responses of the plant using different EWMA filters and Table 10.5 details the key control performance parameters.

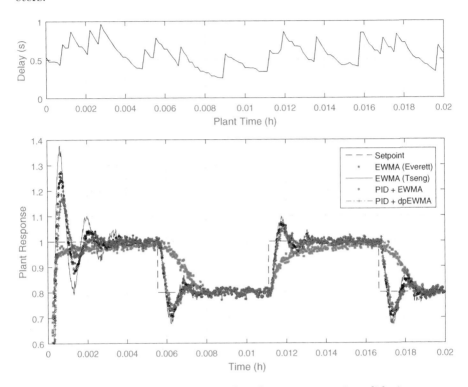

FIGURE 10.7: Comparison of first-order plant responses in validation experiment

TABLE 10.5: Performance comparison of first-order process with presence of noise and model mismatch

Parameter	EWMA (Everett)	EWMA (Tseng)	PID + EWMA	PID + dpEWMA
IAE	155.6552	154.7404	136.5954	130.0592
Rise Time (s)	0.5005	0.6885	0.7853	0.8874
Settling Time (s)	70.56	71.64	71.64	71.64
Overshoot (%)	26.3248	73.3134	59.6959	48.6058

From the figure, it can be seen that Everett's method results in the least percent overshoot, however, in return, it is much slower when responding to process setpoint change. This is an unwanted behavior especially in industrial process plants where fast adaptation to setpoint change is desired. Differently, Tseng's method has the highest percent overshoot although it responds fastest to the setpoint change. Hence it is the key drawback for Tseng's approach being used in first-order fast-sampled plants. Overall the proposed dpEWMA filter has the best performance with lower overshoot, faster response time and the least IAE.

10.2.7.2   Second-order Plant

The plant's responses are presented in Figure 10.8. Different from the first-order plant, the overshoot resulting from the time-varying delay in the second-order plant is higher. Based on the figure, there is insignificant difference between the response of the plant with PID and that with the Kalman filter. Both have the highest overshoot of about 70%. Compared to these responses, the responses from the plant with EWMA filters have lower overshoot. In addition, their oscillations are less aggressive, especially at the high upstream delay of 1 s (about 3 times average value of 0.3 s) at about 0.015 h.

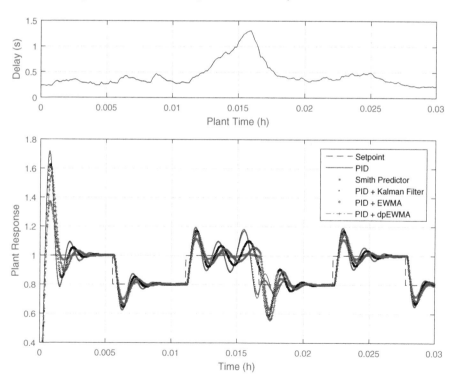

FIGURE 10.8: Second-order plant responses in validation experiment

The effect of high delay on control performance is well represented in the figure, at the second cycle, as well. At 0.0175 h, the responses of EWMA filters are more aggressive because of two main reasons. The first reason is that the setpoint is changed right at the falling edge of the response, thus pulling the signal down further. The second reason is that the plant with PID and Kalman filter has responses slightly slower than with EWMA filters, thus, when the peak delay occurs, its response is already near the new setpoint range. Hence, less effort is needed to keep track of the setpoint. The responses of the plant with PID, Kalman filter and EWMA filters are stabilized at nearly the same moment. This indicates the advantage of EWMA filters compared to the plant with PID only or Kalman filter. For this case, the Smith predictor has the best performance compared to the rest. The detail control performance is outlined in Table 10.6.

TABLE 10.6: Performance validation of second-order plant

Parameter	PID	PID + Smith	PID + Kalman	PID + EWMA	PID + dpEWMA
IAE	100.4294	68.0419	96.5696	92.3555	89.7885
Rise Time (s)	0.5958	0.7059	0.6032	0.6733	0.7605
Settling Time (s)	106.56	105.12	106.56	106.92	105.48
Overshoot (%)	117.6388	70.4991	115.8185	102.7071	90.7255

It is seen from the table that in terms of IAE, the Smith predictor has the lowest value, followed by EWMA filters, then PID with Kalman filter and PID only. Although the PID has the shortest rise time, its settling time is in a similar range with the PID with Kalman filter, and that with the EWMA filter. The dpEWMA filter has slightly longer settling time than the Smith predictor. Both have the lowest settling times. Because of the shortest rise time, PID only results in the highest percent overshoot while the lowest one is for the Smith predictor. Compared to the Kalman filter, EWMA filters result in more than 10% percent overshoot reduction. Overall, the performance of the ideal Smith predictor is the best, followed by EWMA filters and PID with Kalman filter, and PID.

Taking into consideration model mismatch and input noise, the plant responses given different control strategies are presented in Figure 10.9.

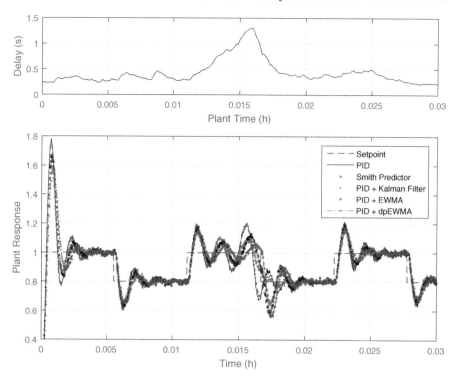

FIGURE 10.9: Second-order plant responses in validation experiment with presence of noise and model mismatch

Here, the responses of the plant have more variation compared to the previous scenario. The peak amplitude is also slightly higher. It also takes longer for the plant to settle especially when oscillation occurs at 0.015 h to 0.02 h. During the oscillation, the plant with PID and Kalman filter has higher amplitude before the step change from 1 to 0.8. Differently, the plant with EWMA, dpEWMA filters and Smith predictor has lower amplitude before the change, and higher amplitude after the change, even though the plant is still stable.

More control performance detail of the plant is presented in Table 10.7. Comparing the result with the previous case and with the case of the first-order plant, it is seen that the second-order plant is less affected by the model mismatch and input noise. As a result, the IAE value is slightly increased. The plant with the Smith predictor is changed the most while that with dpEWMA is changed the least. Overall, the rise time is reduced, while the settling time is increased. This leads to the increment of percent overshoot, from as low as 7% with the dpEWMA filter, to as high as 40% difference with the Smith predictor. Here, again, it is seen that with the help of dpEWMA, the plant's control performance is improved the most.

TABLE 10.7: Performance validation of second-order plant with presence of noise and model mismatch

Parameter	PID	PID + Smith	PID + Kalman	PID + EWMA	PID + dpEWMA
IAE	105.0938	88.9859	101.5434	96.9222	92.8176
Rise Time (s)	0.5699	0.6047	0.5759	0.6415	0.7393
Settling Time (s)	107.64	107.64	107.64	107.64	107.64
Overshoot (%)	128.8057	110.2817	127.1124	111.0935	97.3822

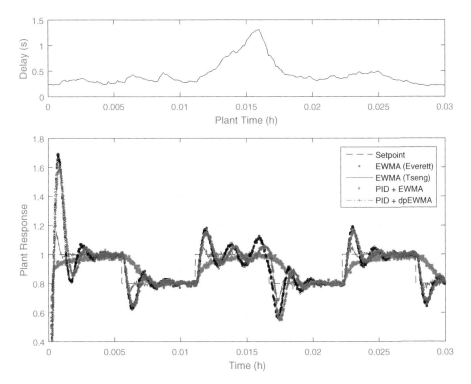

FIGURE 10.10: Comparison of second-order plant responses in validation experiment

In addition, Figure 10.10 and Table 10.8 illustrate and detail the differences between the proposed dpEWMA filters compared to Everett's and Tseng's methods. Unlike the case of the first-order plant, here, in the second-order plant, Everett and Tseng's methods provide better control performance as they are adaptive in nature: lower percent overshoot, faster settling time and lower error rate. This is because for the second order-plant, the delay variation is less and typically of lower magnitude. Although Everett's method has the lowest percent overshoot, its response to setpoint change is still the slowest. Differently, the proposed dpEWMA filter in this case has higher percent overshoot and faster response time. Thus, it can be said that dpEWMA filters result in better control performance for first-order fast-sampled plants, while for second-order plants, Tseng's and Everett's methods are better. For future investigation, the dpEWMA filters can be further developed to be adaptive for better control performance of the second-order plants.

TABLE 10.8: Performance comparison of second-order process with presence of noise and model mismatch

Parameter	EWMA (Everett)	EWMA (Tseng)	PID + EWMA	PID + dpEWMA
IAE	238.9712	147.5053	302.9510	290.6785
Rise Time (s)	0.5034	0.6904	0.6417	0.7192
Settling Time (s)	107.64	107.64	107.64	107.64
Overshoot (%)	24.8938	46.5346	113.6589	99.5307

10.3 Summary

As seen from this chapter, the EWMA filter outperforms the other controllers such as PID, and PID with Kalman filter. This is proof that the advanced EWMA filter is suitable for wireless control applications. To the best of the authors' knowledge, this is among the very first attempts to apply WirelessHART in the wireless control loop of industrial process plants.

11

Conclusions and Future Directions

11.1 Introduction

This chapter concludes the main goal of this book, its research significance, current limitations on the proposed solutions and some recommendations for continuation of this research in future works.

11.2 The dpEWMA Filter and WH-HILS's Significance

This book has introduced a computationally efficient EWMA filter design framework to improve control performance of WNCS. The design takes into consideration both delay and packet dropout caused by contention in the network. It had been proved that the filter is suitable for completely addressing the packet dropout issue which is critical for maintaining the stability of wireless plants. In addition, the demonstration of wired contention between the controller and the node is the evidence of the existence and effect of packet dropout on the network's reliability and plant performance.

Furthermore, this book presents a novel approach to model and simulate WNCS with regard to delay and packet dropout. The relationship between Bernoulli and Markov packet dropout models had been formulated. It can be realized that the Bernoulli model can represent well the randomization of packet dropout in WNCS. However, the Markov model is a better choice for such simulation, as it takes into consideration the relationship between the past and current packets. Because of this relationship, if a packet is transferred successfully, it is likely that the consequent one will be successfully transferred as well. Similarly, if the network is unstable and packet dropout occurs, there is more chance that the later packets will be dropped. This causes a chain effect resulting in a series of dropped packets. Thus, WNCS shall experience severe performance degradation in such cases.

Finally, the development of real-time WirelessHART hardware-in-the-loop simulator (WH-HILS) plays an important role in providing a sufficient real-time framework for validation of the proposed filter design approach. The simulator can be used for assessing not only control performance of various wireless plants, but also for their communications quality.

11.3   The dpEWMA Filter and WH-HILS's Applications

The developed dpEWMA filter and WirelessHART hybrid simulator WH-HILS have wide applications, especially for use in industrial process plants. The following highlights their potential applications.

- Remote filtering configuration for field instrument to improve control performance in subject to communication delay and packet dropout.

- Automatic filter tuning at field instrument to compensate for communication delay and packet dropout in the network.

- Simulation of several types of industrial process plants and controllers in real wireless communication environment.

- Surveying industrial wireless communication of process plants for feasibility study for upgrading wired process plants.

- Controller design of wireless process plants.

- Optimizing industrial wireless networks, controllers of wireless process plants.

- Testing of critical operations of wireless process plants for plant failure preventions without affecting production operation of wireless plants.

- Investigation of issues associated with wireless process plants through deployment of WH-HILS at the field.

For supporting these applications, the following basic configurations of WH-HILS are suggested. This is based on the fact that there are two types of typical control systems categorized based on the number of inputs and outputs, namely single-input-single-output (SISO) and multiple-input-multiple-output (MIMO).

11.3.1   WH-HILS Configuration for SISO System Simulation

For simulation of the SISO system using WH-HILS, a configuration using only one computer shown in Figure 11.1 shall be used. As seen from the figure, the computer is connected to both a wireless gateway and a node, where the communication between them follows WirelessHART standard. In addition, the computer hosts a simulation application that has both Controller and

Plant models. However, the signals from Controller are sent to and received from Control and Feedback points, respectively. These points help to transfer data to the gateway. Then the gateway communicates this information with the wireless node. Information from the node is sent to the Plant module via the Actuator point and the plant sends its output to the wireless node via the Sensor point. Thus, a complete wireless closed-loop feedback control system can be formed.

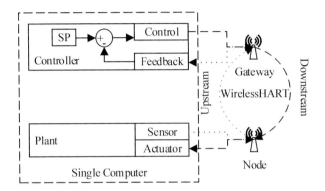

FIGURE 11.1: WH-HILS configuration for SISO system simulations using one computer

The key disadvantage of using a single computer for simulation is the limitation on the distance between the computer and the node or gateway. It is determined by physical communication medium between them. For example, if wires (e.g., USB, Ethernet) are used, the distance shall be limited to several meters. Therefore, it is difficult to bring the wireless node far from the gateway for a more realistic simulation scenario. In order to solve this problem, two computers can be used for simulation. Each holds an application serving as either Controller or Plant. The configuration is shown in Figure 11.2. Since the two computers are separated physically, the distance between the gateway and the node is unlimited.

It can be realized that the limitation of the configurations shown in Figures 11.1 and 11.2 is that the simulator can only simulate one SISO system at a time. In order to overcome this drawback, configurations using one computer and two computers shown in Figures 11.3 and 11.4, respectively, shall be used.

The key difference between the configurations for simulation of one SISO system and those for simulation of two SISO systems is at the additional multiplexer (Mux) and demultiplexer (Demux). These are used to compose and discompose signal components from the Controllers i,j and from the Plants i,j, respectively. For extension of the simulation capability, more Plant and Controller models can be added. However, it should be noted that the more

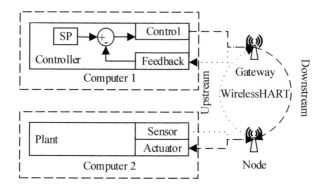

FIGURE 11.2: WH-HILS configuration for SISO system simulations using two computers

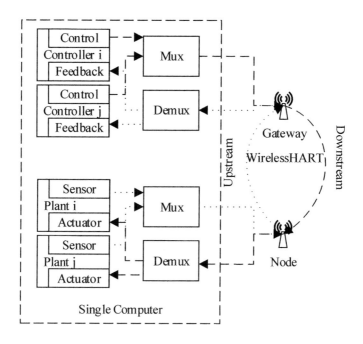

FIGURE 11.3: WH-HILS configuration for two SISO system simulations using one computer

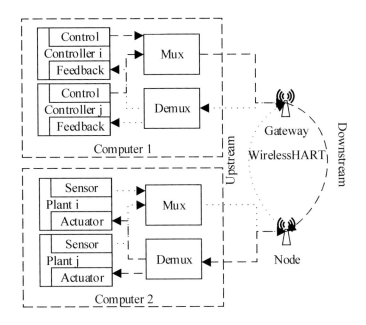

FIGURE 11.4: WH-HILS configuration for two SISO system simulations using two computers

models used in a computer, the more complicated simulations will be. Thus, it will require a computer with higher simulation and computational capability (e.g., higher specifications).

11.3.2 WH-HILS Configuration for MIMO System Simulation

By extending configurations for WH-HILS to simulate SISO systems, it is possible to simulate both single MIMO systems and multiple MIMO systems using one computer, two computers or more. For demonstration, configurations to simulate MIMO systems using one and two computers are shown in Figures 11.5 and 11.6. And configurations for simulation of two MIMO systems using one and two computers are presented in Figures 11.7 and 11.8, correspondingly.

As noticed from these figures, different from SISO cases, for MIMO scenarios, it is important to use different wireless nodes for simulation of separated plants. This is to guarantee that the simulation results represent well the nature of communication between the plants and the gateway using unique wireless modules. Additionally, with MIMO simulation, the foreseen key challenges will be the efficient scheduling mechanism for transferring data between the gateway and the wireless nodes.

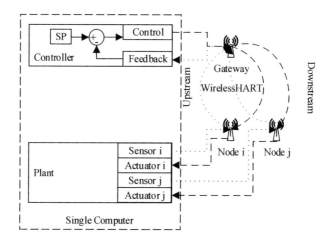

FIGURE 11.5: WH-HILS configuration for one MIMO system simulation using one computer

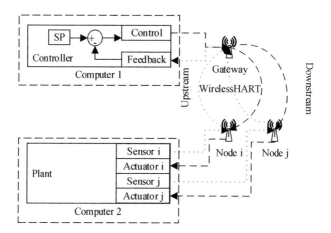

FIGURE 11.6: WH-HILS configuration for one MIMO system simulation using two computers

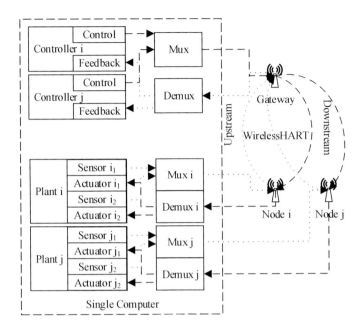

FIGURE 11.7: WH-HILS configuration for two MIMO system simulations using one computer

11.4 The Research Limitations

In this research, the proposed dpEWMA filter has two interrelated parameters which are updateable via control signal sent from the controller. This configuration is useful for fast sampling signals such as those on the control path. However, there is another application of the filter for use in slower sampling signals such as those on the feedback path (see Figure 11.9). In this case, filter weight applied on the most recent signal shall be outdated. This means, if the filter is applied for feedback path, more weights should be used to place sufficient consideration on not only the most recent signal, but also a series of most recent signals. This shall then become a multi-weight moving average filter problem which is worth investigating in the future. The simplification of the filter design relies on its low order, i.e., first-order filter. This shall not provide the best performance to some particular applications that require a higher-order filter. Therefore, higher-order filter design should be considered in future works. In addition, the filter is a slightly slower than PID in response to setpoint change and can result in slightly higher percent overshoot

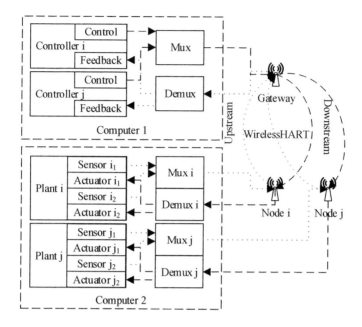

FIGURE 11.8: WH-HILS configuration for two MIMO system simulations using two computers

on control performance. However, with proper configuration, it is effective for addressing both delay and packet dropout issues during setpoint change with percent overshoot close to or smaller than that of a plant with only PID.

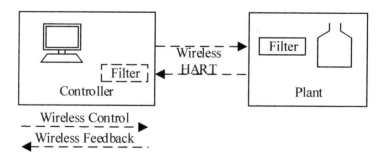

FIGURE 11.9: Possible filter's location in WNCS

11.5   Recommendations for Future Investigations

It is recommended that future work direction should take into account the followings areas:

- Further improvement on the dpEWMA filtering technique for industrial process plants with online tuning, network wide optimization. Online tuning will help to resolve the issue of a fixed update weight. Thus the process plant does not need to preset a suitable value of the weight over a long period of operation. Instead, any movement in the process performance can result in a correspondingly updated weight. This makes the performance of the filter better and the filter can adapt to the change in process operation. The application of the filter on multiple interconnected process plants will require optimization of the set of filter weights. This is because the filter is designed for dedicated control loops, and when multiple loops are connected, the correlation among the weights shall be significant. Therefore, optimization is needed for multiple process tuned weights.

- Smart switching mechanism for a dpEWMA filter to switch between partial and full configurations for further improvement in performance of wireless plants. The switching of the filter from partial to full configuration (for a particular case) helps to maintain performance of the control systems as at different configurations, the filters will have their own advantages such as suitability for handling delay, packet dropout, reducing the effect of step change, maintaining the signal's stability thus the overall system's performance.

- Investigation of the suitability of higher-order dpEWMA filters, such as the second-order one for industrial wireless process plants. For dedicated process applications that require higher-order filters, this is important as it can provide the best performance for the process plant rather than using the same weight for two cascaded first-order filters.

- Investigation of the combination of dpEWMA filtering with other delay and packet dropout compensation techniques for better wireless control performance. Combination of different compensation techniques will help to provide a robust solution for a complex system with multiple process loops as in WHNCS. In addition, by hybridizing or combining different techniques, the best of each can be harnessed to offer superior performance over each individual one.

Bibliography

[1] T. Blevins, D. Chen, M. Nixon, and W. Wojsznis, *Wireless Control Foundation, Continuous and Discrete Control for the Process Industry*, 4th ed. International Society of Automation, 2015.

[2] T. D. Chung, R. B. Ibrahim, V. S. Asirvadam, N. B. Saad, and S. M. Hassan, "Adopting EWMA filter on a fast sampling wired link contention in WirelessHART control system," *IEEE Transactions on Instrumentation and Measurement*, vol. 65, no. 4, pp. 836–845, April 2016.

[3] T. Blevins and M. Nixon, *Control Loop Foundation, Batch and Continuous Processes*. International Society of Automation, 2011.

[4] G. Huang, D. Akopian, and C. L. P. Chen, "Measurement and characterization of channel delays for broadband power line communications," *IEEE Transactions on Instrumentation and Measurement*, vol. 63, no. 11, pp. 2583–2590, Nov 2014.

[5] J. Powell and H. Vandelinde, *Catching the Process Fieldbus, An Introduction to PROFIBUS for Process Automation*. Momentum Press, 2013.

[6] S. K. Sen, *Fieldbus and Networking in Process Automation*. CRC Press, 2014.

[7] H. C. Foundation, *WirelessHART, The First Simple, Reliable and Secure Wireless Standard for Process Monitoring and Control*, 2nd ed. HART Commun. Found., 2009.

[8] FieldComm Group (2017) [Online]. https://fieldcommgroup.org.

[9] A. N. Kim, F. Hekland, S. Petersen, and P. Doyle, "When HART goes wireless: Understanding and implementing the WirelessHART standard," in *2008 IEEE International Conference on Emerging Technologies and Factory Automation*, Sept 2008, pp. 899–907.

[10] IEC, *Industrial Communication Networks, Wireless Communication Network and Communication Profiles, WirelessHART*, 1st ed. International Electrotechnical Commission, 2010.

[11] G. Huang, D. Akopian, and C. L. P. Chen, "Measurement and modeling of network delays for MS-based A-GPS assistance delivery," *IEEE Transactions on Instrumentation and Measurement*, vol. 63, no. 8, pp. 1896–1906, Aug 2014.

[12] I. Al-Anbagi, M. Erol-Kantarci, and H. T. Mouftah, "Delay-aware medium access schemes for WSN-based partial discharge measurement," *IEEE Transactions on Instrumentation and Measurement*, vol. 63, no. 12, pp. 3045–3057, Dec 2014.

[13] J. Fabini and M. Abmayer, "Delay measurement methodology revisited: Time-slotted randomness cancellation," *IEEE Transactions on Instrumentation and Measurement*, vol. 62, no. 10, pp. 2839–2848, Oct 2013.

[14] C. Gomez and J. Paradells, "Wireless home automation networks: A survey of architectures and technologies," *IEEE Communications Magazine*, vol. 48, no. 6, pp. 92–101, June 2010.

[15] M. Nixon, *A Comparison of WirelessHART and ISA 100.11a.* Emerson Process Management, 2012.

[16] A. Saifullah, Y. Xu, C. Lu, and Y. Chen, "End-to-end communication delay analysis in industrial wireless networks," *IEEE Transactions on Computers*, vol. 64, no. 5, pp. 1361–1374, May 2015.

[17] F. Lamonaca, A. Gasparri, E. Garone, and D. Grimaldi, "Clock synchronization in wireless sensor network with selective convergence rate for event driven measurement applications," *IEEE Transactions on Instrumentation and Measurement*, vol. 63, no. 9, pp. 2279–2287, Sept 2014.

[18] P. Zand, E. Mathews, P. Havinga, S. Stojanovski, E. Sisinni, and P. Ferrari, "Implementation of WirelessHART in the NS-2 simulator and validation of its correctness," *Sensors*, vol. 14, no. 5, p. 8633, 2014.

[19] T. E. Abrudan, A. Haghparast, and V. Koivunen, "Time synchronization and ranging in OFDM systems using time-reversal," *IEEE Transactions on Instrumentation and Measurement*, vol. 62, no. 12, pp. 3276–3290, Dec 2013.

[20] T. D. Chung, R. B. Ibrahim, V. S. Asirvadam, N. B. Saad, and S. M. Hassan, "Simulation of WirelessHART networked control system with packet dropout," in *Control Conference (ASCC), 2015 10th Asian*, May 2015, pp. 1–6.

[21] F. Nawaz and V. Jeoti, "Performance assessment of WirelessHART technology for its implementation in dense reader environment," *Computing*, vol. 98, no. 3, pp. 257–277, 2016.

[22] I. Yaqoob, E. Ahmed, I. A. T. Hashem, A. I. A. Ahmed, A. Gani, M. Imran, and M. Guizani, "Internet of things architecture: Recent advances, taxonomy, requirements, and open challenges," *IEEE Wireless Communications*, vol. 24, no. 3, pp. 10–16, 2017.

[23] A. A. Khan, M. H. Rehmani, and A. Rachedi, "Cognitive-radio-based internet of things: Applications, architectures, spectrum related functionalities, and future research directions," *IEEE Wireless Communications*, vol. 24, no. 3, pp. 17–25, 2017.

[24] S. Rani, S. H. Ahmed, R. Talwar, J. Malhotra, and H. Song, "IoMT: A reliable cross layer protocol for internet of multimedia things," *IEEE Internet of Things Journal*, vol. 4, no. 3, pp. 832–839, June 2017.

[25] J. H. Lee and H. Kim, "Security and privacy challenges in the internet of things [security and privacy matters]," *IEEE Consumer Electronics Magazine*, vol. 6, no. 3, pp. 134–136, July 2017.

[26] N. Kaur and S. K. Sood, "An energy-efficient architecture for the internet of things (IoT)," *IEEE Systems Journal*, vol. 11, no. 2, pp. 796–805, June 2017.

[27] W. Yang, M. Wang, J. Zhang, J. Zou, M. Hua, T. Xia, and X. You, "Narrowband wireless access for low-power massive internet of things: A bandwidth perspective," *IEEE Wireless Communications*, vol. 24, no. 3, pp. 138–145, 2017.

[28] F. K. Shaikh, S. Zeadally, and E. Exposito, "Enabling technologies for green internet of things," *IEEE Systems Journal*, vol. 11, no. 2, pp. 983–994, June 2017.

[29] K. Wang, Y. Wang, Y. Sun, S. Guo, and J. Wu, "Green industrial internet of things architecture: An energy-efficient perspective," *IEEE Communications Magazine*, vol. 54, no. 12, pp. 48–54, December 2016.

[30] S. Mumtaz, A. Alsohaily, Z. Pang, A. Rayes, K. F. Tsang, and J. Rodriguez, "Massive internet of things for industrial applications: Addressing wireless IIoT connectivity challenges and ecosystem fragmentation," *IEEE Industrial Electronics Magazine*, vol. 11, no. 1, pp. 28–33, March 2017.

[31] I. Muller, J. C. Netto, and C. E. Pereira, "WirelessHART field devices," *IEEE Instrumentation Measurement Magazine*, vol. 14, no. 6, pp. 20–25, December 2011.

[32] J. Song, S. Han, A. Mok, D. Chen, M. Lucas, M. Nixon, and W. Pratt, "WirelessHART: Applying wireless technology in real-time industrial process control," in *Real-Time and Embedded Technology and Applications Symposium, 2008. RTAS '08. IEEE*, April 2008, pp. 377–386.

[33] C. Shell, J. Henderson, H. Verra, and J. Dyer, "Implementation of a wireless battery management system (WBMS)," in *2015 IEEE International Instrumentation and Measurement Technology Conference (I2MTC) Proceedings*, May 2015, pp. 1954–1959.

[34] ZigBee Alliance (2017) [Online]. http://www.zigbee.org.

[35] T. Nhon and D.-S. Kim, "Real-time message scheduling for ISA100.11a networks," *Computer Standards & Interfaces*, vol. 37, pp. 73–79, 2015.

[36] J. K. Ariza, I. Muller, J. M. Winter, J. C. Netto, C. A. Pereira, and V. J. Brusamarello, "WirelessHART localization algorithm," in *2014 12th IEEE International Conference on Industrial Informatics (INDIN)*, July 2014, pp. 690–695.

[37] I. Muller, J. M. Winter, V. Brusamarello, C. E. Pereira, and J. C. Netto, "Algorithm for estimation of energy consumption of industrial wireless sensor networks nodes," in *2014 IEEE International Instrumentation and Measurement Technology Conference (I2MTC) Proceedings*, May 2014, pp. 440–444.

[38] M. Nobre, I. Silva, and L. A. Guedes, "Routing and scheduling algorithms for WirelessHART networks: A survey," *Sensors*, vol. 15, no. 5, p. 9703, 2015.

[39] Q. Huang, A. Sikora, V. F. Groza, and P. Zand, "Simulation and analysis of WirelessHART nodes for real-time actuator application," in *2014 IEEE International Instrumentation and Measurement Technology Conference (I2MTC) Proceedings*, May 2014, pp. 1590–1594.

[40] G. Habib, N. Haddad, and R. E. Khoury, "Case study: WirelessHART vs ZigBee network," in *Technological Advances in Electrical, Electronics and Computer Engineering (TAEECE), 2015 Third International Conference on*, April 2015, pp. 135–138.

[41] M. Fang, Y. Sun, X. Zhang, and J. Sun, "ELOTS: Energy-efficient local optimization time synchronization algorithm for WirelessHART networks," in *Systems Conference (SysCon), 2014 8th Annual IEEE*, March 2014, pp. 402–406.

[42] A. Ulusoy, O. Gurbuz, and A. Onat, "Wireless model-based predictive networked control system over cooperative wireless network," *IEEE Transactions on Industrial Informatics*, vol. 7, no. 1, pp. 41–51, Feb 2011.

[43] M. D. Natale, H. Zeng, P. Giusto, and A. Ghosal, *Understanding and Using the Controller Area Network Communication Protocol, Theory and Practice*. Springer-Verlag New York, 2012.

[44] I. Glaropoulos, M. Lagan, V. Fodor, and C. Petrioli, "Energy efficient COGnitive-MAC for sensor networks under WLAN co-existence," *IEEE Transactions on Wireless Communications*, vol. 14, no. 7, pp. 4075–4089, July 2015.

[45] D. Taniuchi, X. Liu, D. Nakai, and T. Maekawa, "Spring model based collaborative indoor position estimation with neighbor mobile devices," *IEEE Journal of Selected Topics in Signal Processing*, vol. 9, no. 2, pp. 268–277, March 2015.

[46] Y. A. Millan, F. Vargas, F. Molano, and E. Mojica, "A wireless networked control systems review," in *Robotics Symposium, 2011 IEEE IX Latin American and IEEE Colombian Conference on Automatic Control and Industry Applications (LARC)*, Oct 2011, pp. 1–6.

[47] I. Sigma Designs. (2016) Z-Wave. [Online]. Available: http://www.z-wave.com.

[48] B. SIG, *Specification of the Bluetooth System*, 4th ed. Bluetooth SIG, 2014.

[49] B. Wu, H. Lin, and M. Lemmon, "Stability analysis for wireless networked control system in unslotted IEEE 802.15.4 protocol," in *11th IEEE International Conference on Control Automation (ICCA)*, June 2014, pp. 1084–1089.

[50] S. Petersen and S. Carlsen, "WirelessHART versus ISA100.11a: The format war hits the factory floor," *IEEE Industrial Electronics Magazine*, vol. 5, no. 4, pp. 23–34, Dec 2011.

[51] P. Ferrari, A. Flammini, M. Rizzi, and E. Sisinni, "Improving simulation of wireless networked control systems based on WirelessHART," *Computer Standards & Interfaces*, vol. 35, no. 6, pp. 605–615, 2013.

[52] D. Chen, M. Nixon, S. Han, A. K. Mok, and X. Zhu, "WirelessHART and IEEE 802.15.4e," in *Industrial Technology (ICIT), 2014 IEEE International Conference on*, Feb 2014, pp. 760–765.

[53] S. Raza, A. Slabbert, T. Voigt, and K. Landerns, "Security considerations for the WirelessHART protocol," in *2009 IEEE Conference on Emerging Technologies Factory Automation*, Sept 2009, pp. 1–8.

[54] E. P. Management, *Smart Wireless THUM^{TM} Adapter*. Emerson Process Management, 2014.

[55] ——, *System Engineering Guidelines IEC 62591 WirelessHART*, 4th ed. Emerson Process Management, 2014.

[56] T. D. Chung, R. Ibrahim, V. S. Asirvadam, N. Saad, and S. M. Hassan, "Energy consumption analysis of WirelessHART adaptor for industrial wireless sensor actuator network," *Procedia Computer Science*, vol. 105, pp. 227–234, 2017.

[57] H. Abdullah, R. Ibrahim, S. M. Hassan, and T. D. Chung, "Filtered feedback PID control for WirelessHART networked plant," in *2016 6th International Conference on Intelligent and Advanced Systems (ICIAS)*, Aug 2016, pp. 1–5.

[58] X. Jin, J. Wang, and P. Zeng, "End-to-end delay analysis for mixed-criticality WirelessHART networks," *IEEE/CAA Journal of Automatica Sinica*, vol. 2, no. 3, pp. 282–289, July 2015.

[59] D. E. Seborg, T. F. Edgar, and D. A. Mellichamp, *Process Dynamics and Control*, 2nd ed.   John Wiley & Sons, Inc., 2004.

[60] A. Cervin, D. Henriksson, B. Lincoln, J. Eker, and K. E. Arzen, "How does control timing affect performance? Analysis and simulation of timing using jitterbug and truetime," *IEEE Control Systems*, vol. 23, no. 3, pp. 16–30, June 2003.

[61] M. D. Biasi, C. Snickars, K. Landernas, and A. J. Isaksson, "Simulation of process control with WirelessHART networks subject to packet losses," in *2008 IEEE International Conference on Automation Science and Engineering*, Aug 2008, pp. 548–553.

[62] M. D. Biasi, C. Snickars, K. Landernas, and A. Isaksson, "Simulation of process control with WirelessHART networks subject to clock drift," in *2008 32nd Annual IEEE International Computer Software and Applications Conference*, July 2008, pp. 1355–1360.

[63] C. M. D. Dominicis, P. Ferrari, A. Flammini, E. Sisinni, M. Bertocco, G. Giorgi, C. Narduzzi, and F. Tramarin, "Investigating WirelessHART coexistence issues through a specifically designed simulator," in *2009 IEEE Instrumentation and Measurement Technology Conference*, May 2009, pp. 1085–1090.

[64] I. Konovalov, J. Neander, M. Gidlund, F. Osterlind, and T. Voigt, "Evaluation of WirelessHART enabled devices in a controlled simulation environment," in *2011 IEEE International Symposium on Industrial Electronics*, June 2011, pp. 2009–2014.

[65] M. Nobre, I. Silva, L. A. Guedes, and P. Portugal, "Towards a WirelessHART module for the ns-3 simulator," in *2010 IEEE 15th Conference on Emerging Technologies Factory Automation (ETFA 2010)*, Sept 2010, pp. 1–4.

[66] G. Gao, H. Zhang, and L. Li, "An OPNET-based simulation approach for the deployment of WirelessHART," in *Fuzzy Systems and Knowledge Discovery (FSKD), 2012 9th International Conference on*, May 2012, pp. 2120–2124.

[67] M. Lobbers and D. Willkomm. (2017) Mobility framework for OMNeT++. [Online]. http://mobility-fw.sourceforge.net.

[68] P. Zand, A. Dilo, and P. Havinga, "Implementation of WirelessHART in NS-2 simulator," in *Proceedings of 2012 IEEE 17th International Conference on Emerging Technologies Factory Automation (ETFA 2012)*, Sept 2012, pp. 1–8.

[69] Thingsquare. (2016) Contiki: The open source OS for the Internet of Things. [Online]. Available: http://www.contiki-os.org.

[70] S. C. Wang and Y. H. Liu, "A PSO-based fuzzy-controlled searching for the optimal charge pattern of Li-Ion batteries," *IEEE Transactions on Industrial Electronics*, vol. 62, no. 5, pp. 2983–2993, May 2015.

[71] N. R. Tavana and V. Dinavahi, "A general framework for FPGA-based real-time emulation of electrical machines for HIL applications," *IEEE Transactions on Industrial Electronics*, vol. 62, no. 4, pp. 2041–2053, April 2015.

[72] B. Palmintier, B. Lundstrom, S. Chakraborty, T. Williams, K. Schneider, and D. Chassin, "A power hardware-in-the-loop platform with remote distribution circuit cosimulation," *IEEE Transactions on Industrial Electronics*, vol. 62, no. 4, pp. 2236–2245, April 2015.

[73] J. P. F. Trovo, V. D. N. Santos, C. H. Antunes, P. G. Pereirinha, and H. M. Jorge, "A real-time energy management architecture for multi-source electric vehicles," *IEEE Transactions on Industrial Electronics*, vol. 62, no. 5, pp. 3223–3233, May 2015.

[74] O. Knig, C. Hametner, G. Prochart, and S. Jakubek, "Battery emulation for power-HIL using local model networks and robust impedance control," *IEEE Transactions on Industrial Electronics*, vol. 61, no. 2, pp. 943–955, Feb 2014.

[75] T. Ould-Bachir, H. F. Blanchette, and K. Al-Haddad, "A network tearing technique for FPGA-based real-time simulation of power converters," *IEEE Transactions on Industrial Electronics*, vol. 62, no. 6, pp. 3409–3418, June 2015.

[76] C. S. Edrington, M. Steurer, J. Langston, T. El-Mezyani, and K. Schoder, "Role of power hardware in the loop in modeling and simulation for experimentation in power and energy systems," *Proceedings of the IEEE*, vol. 103, no. 12, pp. 2401–2409, Dec 2015.

[77] M. Rezkallah, A. Hamadi, A. Chandra, and B. Singh, "Real-time HIL implementation of sliding mode control for standalone system based on PV array without using dumpload," *IEEE Transactions on Sustainable Energy*, vol. 6, no. 4, pp. 1389–1398, Oct 2015.

[78] P. C. Kotsampopoulos, F. Lehfuss, G. F. Lauss, B. Bletterie, and N. D. Hatziargyriou, "The limitations of digital simulation and the advantages of PHIL testing in studying distributed generation provision of ancillary services," *IEEE Transactions on Industrial Electronics*, vol. 62, no. 9, pp. 5502–5515, Sept 2015.

[79] G. F. Lauss, M. O. Faruque, K. Schoder, C. Dufour, A. Viehweider, and J. Langston, "Characteristics and design of power hardware-in-the-loop simulations for electrical power systems," *IEEE Transactions on Industrial Electronics*, vol. 63, no. 1, pp. 406–417, Jan 2016.

[80] C. E. Agero, N. Koenig, I. Chen, H. Boyer, S. Peters, J. Hsu, B. Gerkey, S. Paepcke, J. L. Rivero, J. Manzo, E. Krotkov, and G. Pratt, "Inside the virtual robotics challenge: Simulating real-time robotic disaster response," *IEEE Transactions on Automation Science and Engineering*, vol. 12, no. 2, pp. 494–506, April 2015.

[81] C. Wei and D. Soffker, "Optimization strategy for PID-controller design of AMB rotor systems," *IEEE Transactions on Control Systems Technology*, vol. 24, no. 3, pp. 788–803, May 2016.

[82] B. Martinez, X. Vilajosana, F. Chraim, I. Vilajosana, and K. S. J. Pister, "When scavengers meet industrial wireless," *IEEE Transactions on Industrial Electronics*, vol. 62, no. 5, pp. 2994–3003, May 2015.

[83] H. Zhang, P. Soldati, and M. Johansson, "Performance bounds and latency-optimal scheduling for convergecast in WirelessHART networks," *IEEE Transactions on Wireless Communications*, vol. 12, no. 6, pp. 2688–2696, June 2013.

[84] M. Nobre, I. Silva, and L. A. Guedes, "Performance evaluation of WirelessHART networks using a new network simulator 3 module," *Computers & Electrical Engineering*, vol. 41, pp. 325 – 341, 2015.

[85] F. Cacace, F. Conte, and A. Germani, "Filtering continuous-time linear systems with time-varying measurement delay," *IEEE Transactions on Automatic Control*, vol. 60, no. 5, pp. 1368–1373, May 2015.

[86] R. Teo, D. M. Stipanovic, and C. J. Tomlin, "Decentralized spacing control of a string of multiple vehicles over lossy datalinks," *IEEE Transactions on Control Systems Technology*, vol. 18, no. 2, pp. 469–473, March 2010.

[87] Y. Bolea, V. Puig, and J. Blesa, "Gain-scheduled smith predictor PID-based LPV controller for open-flow canal control," *IEEE Transactions on Control Systems Technology*, vol. 22, no. 2, pp. 468–477, March 2014.

[88] D. E. Quevedo, J. Ostergaard, and A. Ahln, "Power control and coding formulation for state estimation with wireless sensors," *IEEE Transactions on Control Systems Technology*, vol. 22, no. 2, pp. 413–427, March 2014.

[89] J. Akerberg, F. Reichenbach, and M. Bjrkman, "Enabling safety-critical wireless communication using WirelessHART and PROFIsafe," in *Emerging Technologies and Factory Automation (ETFA), 2010 IEEE Conference on*, Sept 2010, pp. 1–8.

[90] X. Zhu, T. Lin, S. Han, A. Mok, D. Chen, M. Nixon, and E. Rotvold, "Measuring WirelessHART against wired fieldbus for control," in *IEEE 10th International Conference on Industrial Informatics*, July 2012, pp. 270–275.

[91] P. Ferrari, A. Flammini, S. Rinaldi, and E. Sisinni, "Performance assessment of a WirelessHART network in a real-world testbed," in *Instrumentation and Measurement Technology Conference (I2MTC), 2012 IEEE International*, May 2012, pp. 953–957.

[92] A. Saifullah, Y. Xu, C. Lu, and Y. Chen, "End-to-end delay analysis for fixed priority scheduling in WirelessHART networks," in *2011 17th IEEE Real-Time and Embedded Technology and Applications Symposium*, April 2011, pp. 13–22.

[93] A. Remke and X. Wu, "WirelessHART modeling and performance evaluation," in *2013 43rd Annual IEEE/IFIP International Conference on Dependable Systems and Networks (DSN)*, June 2013, pp. 1–12.

[94] H. S. Miya, R. B. Ibrahim, N. B. Saad, V. S. Asirvadam, and T. D. Chung, "WirelessHART process control with Smith predictor compensator," *Advanced Science Letters*, vol. 22, no. 10, 2016.

[95] S. M. Hassan, R. Ibrahim, N. Saad, V. S. Asirvadam, and T. D. Chung, "Predictive PI controller for wireless control system with variable network delay and disturbance," in *2016 2nd IEEE International Symposium on Robotics and Manufacturing Automation (ROMA)*, Sept 2016, pp. 1–6.

[96] A. Leva, F. Terraneo, L. Rinaldi, A. V. Papadopoulos, and M. Maggio, "High-precision low-power wireless nodes' synchronization via decentralized control," *IEEE Transactions on Control Systems Technology*, vol. 24, no. 4, pp. 1279–1293, July 2016.

[97] L. Technology, *SmartMesh WirelessHART Application Notes*, 2014.

[98] S. Dey, A. Chiuso, and L. Schenato, "Remote estimation with noisy measurements subject to packet loss and quantization noise," *IEEE Transactions on Control of Network Systems*, vol. 1, no. 3, pp. 204–217, Sept 2014.

[99] C.-C. Hua, Y. Zheng, and X.-P. Guan, "Modeling and control for wireless networked control system," *International Journal of Automation and Computing*, vol. 8, no. 3, pp. 357–363, 2011.

[100] U. Dohare, D. K. Lobiyal, and S. Kumar, "Energy balanced model for lifetime maximization in randomly distributed wireless sensor networks," *Wireless Personal Communications*, vol. 78, no. 1, pp. 407–428, 2014.

[101] T. Lu and J. Zhu, "Genetic algorithm for energy-efficient QoS multicast routing," *IEEE Communications Letters*, vol. 17, no. 1, pp. 31–34, January 2013.

[102] "Ad-ATMA: An efficient MAC protocol for wireless sensor and ad hoc networks," *Procedia Computer Science*, vol. 52, pp. 484–491, 2015.

[103] V. K. Verma, S. Singh, and N. P. Pathak, "Optimized battery models observations for static, distance vector and on-demand based routing protocols over 802.11 enabled wireless sensor networks," *Wireless Personal Communications*, vol. 81, no. 2, pp. 503–517, 2015.

[104] R. Hou and M. Zheng, "Packet-based nonlinear battery energy consumption optimizing for WSNs nodes," *IEICE Electronics Express*, vol. 11, no. 9, pp. 20 140 167–20 140 167, 2014.

[105] R. Hou, Y. Chen, and G. Xing, "Path selection for optimizing nonlinear battery energy consumption," *IEICE Electronics Express*, vol. 10, no. 22, pp. 20 130 689–20 130 689, 2013.

[106] N. Semiconductor, *nRF24L01+ Single Chip 2.4 GHz Transceiver*, Revision 1.0 ed.   Nordic Semiconductor, 2008.

[107] STMicroelectronics, *STM32L051x6 STM32L051x8 - Access line ultra-low-power 32-bit MCU ARM®-based Cortex®-M0+, up to 64 KB Flash, 8 KB SRAM, 2 KB EEPROM, ADC*, DocID025938 Rev 4 ed. STMicroelectronics, 2014.

[108] I. Maxim Integrated Products, *DS8500 HART Modem*, Rev 1; 2/09 ed. Maxim Integrated Products, Inc., 2009.

[109] Saft, *Saft Lithium Batteries Selector Guide*, 54083-2-0515 ed.   Saft, 2015.

[110] T. Supramaniam, R. Ibrahim, T. D. Chung, and S. M. Hassan, "Towards the development of WirelessHART adaptor for process control applications," in *2016 6th International Conference on Intelligent and Advanced Systems (ICIAS)*, Aug 2016, pp. 1–6.

[111] K. M. S. Thotahewa, J. M. Redout, and M. R. Yuce, "Propagation, power absorption, and temperature analysis of UWB wireless capsule endoscopy devices operating in the human body," *IEEE Transactions on Microwave Theory and Techniques*, vol. 63, no. 11, pp. 3823–3833, Nov 2015.

[112] A. Zajic, *Mobile-to-Mobile Wireless Channels*, 1st ed. Artech House, Inc., 2013.

[113] D. Jaramillo-Ramirez, M. Kountouris, and E. Hardouin, "Coordinated multi-point transmission with imperfect CSI and other-cell interference," *IEEE Transactions on Wireless Communications*, vol. 14, no. 4, pp. 1882–1896, April 2015.

[114] S. H. Song, A. F. Almutairi, and K. B. Letaief, "Outage-capacity based adaptive relaying in LTE-advanced networks," *IEEE Transactions on Wireless Communications*, vol. 12, no. 9, pp. 4778–4787, September 2013.

[115] M. Drieberg, V. S. Asirvadam, and F. C. Zheng, "Accurate delay analysis in prioritised wireless sensor networks for generalized packet arrival," *IEEE Wireless Communications Letters*, vol. 3, no. 2, pp. 205–208, April 2014.

[116] Linear Technology, *DC9007 SmartMesh WirelessHART Starter Kit*, 0813b ed. Linear Technology, 2015.

[117] T. D. Chung, V. S. Asirvadam, R. B. Ibrahim, N. B. Saad, and S. M. Hassan, "Analyzing WirelessHART high-level data link control frame payload," in *2015 IEEE International Conference on Signal and Image Processing Applications (ICSIPA)*, Oct 2015, pp. 27–31.

[118] T. D. Chung, R. Ibrahim, V. S. Asirvadam, N. Saad, and S. M. Hassan, "Latency analysis of WirelessHART control message with variable payload," in *2016 2nd IEEE International Symposium on Robotics and Manufacturing Automation (ROMA)*, Sept 2016, pp. 1–5.

[119] W. Stallings, *Data and Computer Communications*, 8th ed. Pearson Education, Inc., 2007.

[120] T. D. Chung, R. Ibrahim, V. S. Asirvadam, N. Saad, and S. M. Hassan, "Effect of network induced delays on WirelessHART control system," in *2016 6th International Conference on Intelligent and Advanced Systems (ICIAS)*, Aug 2016, pp. 1–5.

[121] B. Chen, L. Yu, and W. A. Zhang, "Robust Kalman filtering for uncertain state delay systems with random observation delays and missing measurements," *IET Control Theory Applications*, vol. 5, no. 17, pp. 1945–1954, Nov 2011.

[122] A. K. Mishra, S. R. Shimjith, T. U. Bhatt, and A. P. Tiwari, "Kalman filter-based dynamic compensator for vanadium self powered neutron detectors," *IEEE Transactions on Nuclear Science*, vol. 61, no. 3, pp. 1360–1368, June 2014.

[123] G. Rigatos, P. Siano, and N. Zervos, "Sensorless control of distributed power generators with the derivative-free nonlinear Kalman filter," *IEEE Transactions on Industrial Electronics*, vol. 61, no. 11, pp. 6369–6382, Nov 2014.

[124] T. Shi, Z. Wang, and C. Xia, "Speed measurement error suppression for PMSM control system using self-adaption Kalman observer," *IEEE Transactions on Industrial Electronics*, vol. 62, no. 5, pp. 2753–2763, May 2015.

[125] C. C. Hsu and C. T. Su, "An adaptive forecast-based chart for non-Gaussian processes monitoring: With application to equipment malfunctions detection in a thermal power plant," *IEEE Transactions on Control Systems Technology*, vol. 19, no. 5, pp. 1245–1250, Sept 2011.

[126] C. T. Chen and Y. C. Chuang, "An intelligent run-to-run control strategy for chemical-mechanical polishing processes," *IEEE Transactions on Semiconductor Manufacturing*, vol. 23, no. 1, pp. 109–120, Feb 2010.

[127] J. Wang, "Properties of EWMA controllers with gain adaptation," *IEEE Transactions on Semiconductor Manufacturing*, vol. 23, no. 2, pp. 159–167, May 2010.

[128] B. Y. Jou, Y. T. Chan, and S. T. Tseng, "Designing a variable EWMA controller for process disturbance subject to linear drift and step changes," *IEEE Transactions on Semiconductor Manufacturing*, vol. 25, no. 4, pp. 614–622, Nov 2012.

[129] E. P. Management, *DeltaV v11 PID Enhancements for Wireless*. Emerson Process Management, 2013.

[130] Z. Sun, J. Chen, and X. Zhu, "Multi-model internal model control applied in temperature reduction system," in *Intelligent Control and Automation (WCICA), 2014 11th World Congress on*, June 2014, pp. 247–250.

[131] C. Othman, I. B. Cheikh, and D. Soudani, "Application of the internal model control method for the stability study of uncertain sampled systems," in *Electrical Sciences and Technologies in Maghreb (CISTEM), 2014 International Conference on*, Nov 2014, pp. 1–7.

[132] C. Xia, Y. Yan, P. Song, and T. Shi, "Voltage disturbance rejection for matrix converter-based PMSM drive system using internal model control," *IEEE Transactions on Industrial Electronics*, vol. 59, no. 1, pp. 361–372, Jan 2012.

[133] S. Li and H. Gu, "Fuzzy adaptive internal model control schemes for PMSM speed-regulation system," *IEEE Transactions on Industrial Informatics*, vol. 8, no. 4, pp. 767–779, Nov 2012.

[134] S. Saxena and Y. V. Hote, "Load frequency control in power systems via internal model control scheme and model-order reduction," *IEEE Transactions on Power Systems*, vol. 28, no. 3, pp. 2749–2757, Aug 2013.

[135] I. Pan, S. Das, and A. Gupta, "Tuning of an optimal fuzzy {PID} controller with stochastic algorithms for networked control systems with random time delay," {*ISA*} *Transactions*, vol. 50, no. 1, pp. 28–36, 2011.

[136] N. L. Diaz, T. Dragievi, J. C. Vasquez, and J. M. Guerrero, "Intelligent distributed generation and storage units for DC microgrids–a new concept on cooperative control without communications beyond droop control," *IEEE Transactions on Smart Grid*, vol. 5, no. 5, pp. 2476–2485, Sept 2014.

[137] S. Lu, P. Zhou, T. Chai, and W. Dai, "Modeling and simulation of whole ball mill grinding plant for integrated control," *IEEE Transactions on Automation Science and Engineering*, vol. 11, no. 4, pp. 1004–1019, Oct 2014.

[138] R. P. Good, D. Pabst, and J. B. Stirton, "Compensating for the initialization and sampling of EWMA run-to-run controlled processes," *IEEE Transactions on Semiconductor Manufacturing*, vol. 23, no. 2, pp. 168–177, May 2010.

[139] Y. Zheng, D. S. H. Wong, Y. W. Wang, and H. Fang, "Takagi-Sugeno model based analysis of EWMA RtR control of batch processes with stochastic metrology delay and mixed products," *IEEE Transactions on Cybernetics*, vol. 44, no. 7, pp. 1155–1168, July 2014.

[140] A. C. Lee, J. H. Horng, T. W. Kuo, and N. H. Chou, "Robustness analysis of mixed product run-to-run control for semiconductor process based on ODOB control structure," *IEEE Transactions on Semiconductor Manufacturing*, vol. 27, no. 2, pp. 212–222, May 2014.

[141] F. Harirchi, T. L. Vincent, A. Subramanian, K. Poolla, and B. Stirton, "On the initialization of threaded run-to-run control of semiconductor manufacturing," *IEEE Transactions on Semiconductor Manufacturing*, vol. 27, no. 4, pp. 515–522, Nov 2014.

[142] M. D. Ma and J. Y. Li, "Improved variable EWMA controller for general ARIMA processes," *IEEE Transactions on Semiconductor Manufacturing*, vol. 28, no. 2, pp. 129–136, May 2015.

[143] C. F. Chien, Y. J. Chen, C. Y. Hsu, and H. K. Wang, "Overlay error compensation using advanced process control with dynamically adjusted proportional-integral R2R controller," *IEEE Transactions on Automation Science and Engineering*, vol. 11, no. 2, pp. 473–484, April 2014.

[144] C. C. Chang, T. H. Pan, D. S. H. Wong, and S. S. Jang, "An adaptive-tuning scheme for g & p EWMA run-to-run control," *IEEE Transactions on Semiconductor Manufacturing*, vol. 25, no. 2, pp. 230–237, May 2012.

[145] F. Auger, M. Hilairet, J. M. Guerrero, E. Monmasson, T. Orlowska-Kowalska, and S. Katsura, "Industrial applications of the Kalman filter: A review," *IEEE Transactions on Industrial Electronics*, vol. 60, no. 12, pp. 5458–5471, Dec 2013.

[146] M. E. Camargo, W. P. Filho, S. L. Russo, and A. I. dos Santos Dullius, "Kalman filter an application in process control," in *Computers and Industrial Engineering (CIE), 2010 40th International Conference on*, July 2010, pp. 1–3.

[147] G. G. Rigatos, "A derivative-free Kalman filtering approach for sensorless control of nonlinear systems," in *2010 IEEE International Symposium on Industrial Electronics*, July 2010, pp. 2049–2054.

[148] O. Aydogmus and M. F. Talu, "Comparison of extended-Kalman- and particle-filter-based sensorless speed control," *IEEE Transactions on Instrumentation and Measurement*, vol. 61, no. 2, pp. 402–410, Feb 2012.

[149] N. K. Quang, N. T. Hieu, and Q. P. Ha, "FPGA-based sensorless pmsm speed control using reduced-order extended Kalman filters," *IEEE Transactions on Industrial Electronics*, vol. 61, no. 12, pp. 6574–6582, Dec 2014.

[150] C. Moon, K. H. Nam, M. K. Jung, C. H. Chae, and Y. A. Kwon, "Sensorless speed control of permanent magnet synchronous motor using unscented Kalman filter," in *SICE Annual Conference (SICE), 2012 Proceedings of*, Aug 2012, pp. 2018–2023.

[151] F. Alonge, T. Cangemi, F. D'Ippolito, A. Fagiolini, and A. Sferlazza, "Convergence analysis of extended Kalman filter for sensorless control of induction motor," *IEEE Transactions on Industrial Electronics*, vol. 62, no. 4, pp. 2341–2352, April 2015.

[152] V. Smidl and Z. Peroutka, "Advantages of square-root extended Kalman filter for sensorless control of AC drives," *IEEE Transactions on Industrial Electronics*, vol. 59, no. 11, pp. 4189–4196, Nov 2012.

[153] F. Deng, J. Chen, and C. Chen, "Adaptive unscented Kalman filter for parameter and state estimation of nonlinear high-speed objects," *Journal of Systems Engineering and Electronics*, vol. 24, no. 4, pp. 655–665, Aug 2013.

[154] M. Partovibakhsh and G. Liu, "Adaptive unscented Kalman filter-based online slip ratio control of wheeled-mobile robot," in *Intelligent Control and Automation (WCICA), 2014 11th World Congress on*, June 2014, pp. 6161–6166.

[155] A. T. Nair, T. K. Radhakrishnan, K. Srinivasan, and S. R. Valsalam, "Kalman filter based state estimation of a thermal power plant," in *Process Automation, Control and Computing (PACC), 2011 International Conference on*, July 2011, pp. 1–5.

[156] E. H. Ha and K. Park, "Kalman filtering in position control using a vision sensor," in *Control Automation and Systems (ICCAS), 2010 International Conference on*, Oct 2010, pp. 1252–1254.

[157] Y. Noda and K. Terashima, "Simplified flow rate estimation by decentralization of Kalman filters in automatic pouring robot," in *SICE Annual Conference (SICE), 2012 Proceedings of*, Aug 2012, pp. 1465–1470.

[158] S. Deshmukh, B. Natarajan, and A. Pahwa, "State estimation and voltage/VAR control in distribution network with intermittent measurements," *IEEE Transactions on Smart Grid*, vol. 5, no. 1, pp. 200–209, Jan 2014.

[159] J. Zhang, G. Welch, G. Bishop, and Z. Huang, "A two-stage Kalman filter approach for robust and real-time power system state estimation," *IEEE Transactions on Sustainable Energy*, vol. 5, no. 2, pp. 629–636, April 2014.

[160] T. Temel and H. Ashrafiuon, "Sliding-mode speed controller for tracking of underactuated surface vessels with extended Kalman filter," *Electronics Letters*, vol. 51, no. 6, pp. 467–469, 2015.

[161] G. A. Susto, A. Beghi, and C. D. Luca, "A predictive maintenance system for epitaxy processes based on filtering and prediction techniques," *IEEE Transactions on Semiconductor Manufacturing*, vol. 25, no. 4, pp. 638–649, Nov 2012.

[162] H. Karimipour and V. Dinavahi, "Extended Kalman filter-based parallel dynamic state estimation," *IEEE Transactions on Smart Grid*, vol. 6, no. 3, pp. 1539–1549, May 2015.

[163] K. Rajawat, E. Dall'Anese, and G. B. Giannakis, "Dynamic network delay cartography," *IEEE Transactions on Information Theory*, vol. 60, no. 5, pp. 2910–2920, May 2014.

[164] M. Ghanbari and M. J. Yazdanpanah, "Delay compensation of tilt sensors based on mems accelerometer using data fusion technique," *IEEE Sensors Journal*, vol. 15, no. 3, pp. 1959–1966, March 2015.

[165] M. Nourian, A. S. Leong, S. Dey, and D. E. Quevedo, "An optimal transmission strategy for Kalman filtering over packet dropping links with imperfect acknowledgements," *IEEE Transactions on Control of Network Systems*, vol. 1, no. 3, pp. 259–271, Sept 2014.

[166] M. Moayedi, Y. K. Foo, and Y. C. Soh, "Adaptive Kalman filtering in networked systems with random sensor delays, multiple packet dropouts and missing measurements," *IEEE Transactions on Signal Processing*, vol. 58, no. 3, pp. 1577–1588, March 2010.

[167] F. V. Lima, M. R. Rajamani, T. A. Soderstrom, and J. B. Rawlings, "Covariance and state estimation of weakly observable systems: Application to polymerization processes," *IEEE Transactions on Control Systems Technology*, vol. 21, no. 4, pp. 1249–1257, July 2013.

[168] A. Denasi, D. Kosti, and H. Nijmeijer, "Time delay compensation in bilateral teleoperations using IMPACT," *IEEE Transactions on Control Systems Technology*, vol. 21, no. 3, pp. 704–715, May 2013.

[169] D. C. Delgado and J. L. Rivera, "Performance study of distributed power control algorithms under time-delays and measurement uncertainty," *IEEE Latin America Transactions*, vol. 11, no. 2, pp. 690–697, March 2013.

[170] M. Bowthorpe, M. Tavakoli, H. Becher, and R. Howe, "Smith predictor-based robot control for ultrasound-guided teleoperated beating-heart surgery," *IEEE Journal of Biomedical and Health Informatics*, vol. 18, no. 1, pp. 157–166, Jan 2014.

[171] C. Guan, P. B. Luh, L. D. Michel, and Z. Chi, "Hybrid Kalman filters for very short-term load forecasting and prediction interval estimation," *IEEE Transactions on Power Systems*, vol. 28, no. 4, pp. 3806–3817, Nov 2013.

[172] N. J. Ploplys, P. A. Kawka, and A. G. Alleyne, "Closed-loop control over wireless networks," *IEEE Control Systems*, vol. 24, no. 3, pp. 58–71, Jun 2004.

[173] D. E. Seborg, T. F. Edgar, and D. A. Mellichamp, *Process Dynamics and Control*, 2nd ed. John Wiley & Sons, Inc., 2004.

[174] M. Cescon and R. Johansson, "Glycemic trend prediction using empirical model identification," in *Decision and Control, 2009 held jointly with the 2009 28th Chinese Control Conference. CDC/CCC 2009. Proceedings of the 48th IEEE Conference on*. IEEE, 2009, pp. 3501–3506.

[175] H. K. Alaei, K. Salahshoor, and H. K. Alaei, "Model predictive control of distillation column based recursive parameter estimation method using HYSYS simulation," in *Intelligent Computing and Cognitive Informatics (ICICCI), 2010 International Conference on*. IEEE, 2010, pp. 308–311.

[176] Z. Li, M. Hayashibe, Q. Zhang, and D. Guiraud, "FES-induced muscular torque prediction with evoked EMG synthesized by NARX-type recurrent neural network," in *2012 IEEE/RSJ International Conference on Intelligent Robots and Systems*. IEEE, 2012, pp. 2198–2203.

[177] N. Rosli, R. Ibrahim, I. Ismail, S. M. Hassan, and T. D. Chung, "Neural network architecture selection for efficient prediction model of gas metering system," in *2016 2nd IEEE International Symposium on Robotics and Manufacturing Automation (ROMA)*, Sept 2016, pp. 1–5.

[178] C. Guan, P. B. Luh, L. D. Michel, and Z. Chi, "Hybrid Kalman filters for very short-term load forecasting and prediction interval estimation," *IEEE Transactions on Power Systems*, vol. 28, no. 4, pp. 3806–3817, 2013.

[179] F. Alonge, T. Cangemi, F. D'Ippolito, A. Fagiolini, and A. Sferlazza, "Convergence analysis of extended Kalman filter for sensorless control of induction motor," *IEEE Transactions on Industrial Electronics*, vol. 62, no. 4, pp. 2341–2352, 2015.

[180] MathWorks, *Simulink Control Design User's Guide*. MathWorks, 2016.

[181] D. E. Seborg, T. F. Edgar, and D. A. Mellichamp, *Process Dynamics and Control*, 2nd ed. John Wiley & Sons, Inc., 2004.

[182] MathWorks, *MATLAB® - External Interfaces*. MathWorks, 2016.

[183] L. Technology, *SmartMesh WirelessHART Manager API Guide*, 2015.

[184] J. E. Everett, "The exponentially weighted moving average applied to the control and monitoring of varying sample sizes," *WIT Transactions on Modelling and Simulation*, vol. 51, pp. 3–13, 2011.

Index